TEST YOUR GENERAL

MATH

KNOWLEDGE

3.14159265

Chris McMullen, Ph.D.

Test Your General Math Knowledge
Chris McMullen, Ph.D.
3.14159265

www.monkeyphysicsblog.wordpress.com
www.improveyourmathfluency.com
www.chrismcmullen.wordpress.com

Zishka Publishing

ISBN: 978-1-941691-68-7

Science & Math > Mathematics
Entertainment > Puzzles > Quizzes
Entertainment > Games > Trivia

CONTENTS

5	88	49
B	**Ra**	**In**
Boron	Radium	Indium
10.8	226.0	114.8

84	74	68
Po	**W**	**Er**
Polonium	Tungsten	Erbium
209.0	183.8	167.3

INTRODUCTION

What is the scope of your knowledge of mathematics? The chapters of this book will test your knowledge of essential topics in a variety of ways.

- Chapter 1 tests your familiarity with common terms.
- Chapter 2 challenges you with numbers.
- Chapter 3 tests your familiarity with symbols.
- Chapter 4 tests your familiarity with formulas.
- Chapter 5 challenges you to figure out mathematical patterns.
- Chapter 6 tests your math skills.
- Chapter 7 involves mathematical principles.
- Chapters 8-9 feature math history and abbreviations.

The questions in each chapter are organized by level and subject. Everybody should be able to enjoy the majority of each chapter. Towards the end of the chapter, some readers may encounter material that is beyond what they have studied. When you score your test, you should not penalize yourself for questions relating to subjects which you have not learned.

The answer to every problem can be found in a key at the back of the book.

Can you learn math by reading this book? Good question. Here are a few suggestions:

- When you finish a chapter, check your answers at the back of the book.
- Spend time thinking about any mistakes that you made. Try to learn from them.
- If you identify a topic of interest or an area that you would like to improve, find a suitable resource to help you learn more about that topic.

1 Test Your Vocabulary

1) To perform the subtraction 54 − 39, we may _____ from the 5, allowing us to first write 14 − 9 and then write 4 − 3.

2) To perform the addition 76 + 28, after we do 6 + 8 = 14, we _____ the 1 to write 1 + 7 + 2. (2 words)

3) First, second, third, fourth, fifth, etc. are called _____ numbers.

4) 2, 3, 5, 7, 11, 13, 17, 19, ... are called _____ numbers.

5) 4, 6, 8, 9, 10, 12, 14, 15, 16, 18, ... are called _____ numbers.

6) An object that is circular or spherical in shape is said to be _____.

7) A _____ is an instrument which pivots on a point and holds a pencil on its second leg. It is used for drawing a circle by hand.

8) A _____ is an instrument in the shape of a semicircle that is used to measure an angle.

9) An _____ is an instrument that consists of beads that can slide along wires. It can be used to count or do arithmetic.

10) A _____ is an instrument that consists of a small piece that slides along a longer piece. Each piece is marked with a scale. It can be used to do arithmetic. (2 words)

11) A _____ is straight, thin, about a foot long, and used to measure length.

12) A _____ is perfectly straight and is used for drawing lines. (2 words)

13) A _____ is a template that is used for drawing smooth curves. (2 words)

14) 1, 1, 2, 3, 5, 8, 13, 21, ... is known as the _____ sequence.

15) 2, 1, 3, 4, 7, 11, 18, 29, ... are known as _____ numbers.

16) _____ represents a value that exceeds any possible bound.

17) _____ represents the tiniest possible amount, approaching a limit of zero.

18) _____ is more than the answer to #17, but less than the answer to #16.

19) _____ is the number for nothing. As a digit, it is called _____.

20) _____ is the number for a single thing. This value is called _____.

21) 1, 2, 3, 4, 5, ... are Arabic _____ and I, II, III, IV, V, ... are the Roman variety.

22) _____ is a fancy word for the verb count.

23) _____ is the entire amount. _____ is the value after deductions.

24) _____ refers to a measurable value. _____ refers to a characteristic.

25) 2, 4, 6, 8, ... are _____ whereas 1, 3, 5, 7, ... are _____.

26) _____ is the attribute of an integer corresponding to the answers to #25.

27) A _____, _____, and _____ are groups of two, three, and four numbers, respectively.

28) _____ means to set two expressions equal to one another.

29) Two _____ quantities are equal in value.

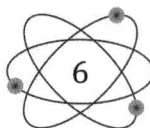

30) $49.95 ÷ 5 is _____ equal to $9.99 and _____ equal to $10.

31) The problem $361 ÷ $8.92 can be _____ by _____ $361 down to $360 and $8.92 up to $9.

32) _____ means to drop extra digits. Examples include changing 37.9 to 37 and 83.6 to 83.

33) Addition, subtraction, multiplication, and division are examples of _____.

34) The +, −, ×, and ÷ symbols are examples of _____.

35) 3! = 3 × 2 × 1 and 4! = 4 × 3 × 2 × 1 are called _____.

36) = is an _____ sign whereas ≠ is an _____ sign.

37) ≈ means _____ (3 words) while ~ means _____ (4 words).

38) < means _____ (2 words) while > means _____ (2 words).

39) ≤ means _____ (5 words) while ≥ means _____ (5 words).

40) 4 and 5 are _____ whereas 9 is the _____ in 4 + 5 = 9.

41) 18 is the _____, 11 is the _____, and 7 is the _____ in 18 − 11 = 7.

42) 5 and 7 are _____ whereas 35 is the _____ in 5 × 7 = 35.

43) 48 is the _____, 12 is the _____, and 4 is the _____ in 48 ÷ 12 = 4.

44) 12, 18, 24, 30, 36, 42, ... are _____ of 6.

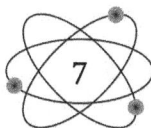

45) When performing the division 20 ÷ 6, there is a _____ equal to 2.

46) 5^2 means five _____ whereas 5^3 means five _____.

47) $\sqrt{7}$ means the _____ of 7. (2 words)

48) $\sqrt[3]{7}$ means the _____ of 7. (2 words)

49) $\frac{3}{4}$ and $\frac{5}{2}$ are examples of _____.

50) $\frac{3}{4}$ is a _____ (2 words) whereas $\frac{5}{2}$ is an _____ (2 words).

51) When we rewrite $\frac{6}{9}$ as $\frac{2}{3}$, we say that $\frac{2}{3}$ is the _____ form.

52) $\frac{5}{8}$ is a _____ (2 words) whereas $\frac{1/6}{3/4}$ is a _____ (2 words).

53) A number of the form $4\frac{2}{5}$ or $12\frac{1}{3}$ is called a _____. (2 words)

54) In $\frac{5}{9}$, the 5 is called the _____ and the 9 is called the _____.

55) A number of the form 0.273 or 8.4 is called a _____.

56) The number $0.\overline{3}$, meaning $0.33333\cdots$ (repeating forever) is called a _____. (2 words)

57) The numbers 3.2×10^4 and 7.59×10^{-6} use _____. (2 words)

58) 42% is a _____, whereas if we say "a _____ of the students," this refers to an unspecified fraction of the students. (Which form of the word we use depends on whether we refer to a precise value, like 42%, or a general unknown amount.)

59) If there are 3 boys for every 4 girls, we can express this with the _____ 3:4.

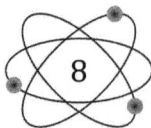

60) When we say that doubling distance will double time, we are making a _____.

61) A number is _____ if greater than zero and _____ if less than zero.

62) A _____ number could be either answer to #61.

63) A _____ number is greater than or equal to zero.

64) The _____ is what distinguishes -18 from 18.

65) An _____ is a number that doesn't have a fractional part, like -3, 0, or 21.

66) A _____ number is a number that doesn't have a fractional part. Some people consider it the same as the answer to #65. Others require it not to be negative.

67) A _____ number is a number that can be counted. Some people require it to be positive. Others allow it to include the zero. (That is, precise usage varies.)

68) $\frac{1}{6}$, 2, and 3.5 are examples of _____ numbers, whereas $\sqrt{5}$ is _____.

69) $\frac{1}{6}$, 2, 3.5, and $\sqrt{5}$ are all _____ numbers, whereas $\sqrt{-5}$ is _____.

70) A number of the form $3 + 2i$, where i represents $\sqrt{-1}$, is said to be _____.

71) The . that appears in 1.73 is called a _____. (2 words)

72) The numbers 6, 3, 8, 5, and 2 appearing in 63.852 are called _____.

73) The _____ (2 words) of the 7 is different in 70, 0.7, and 0.007.

74) In 9.164, the 9 is the _____ (2 words) and 4 is in the _____ (2 words).

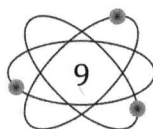

75) The number 36.597 has three _____. (2 words)

76) The number 0.00000415 has three _____. (2 words)

77) The number 1.700 ends with two _____. (2 words)

78) In $3 \times (2 + 5)$, the $2 + 5$ is enclosed in _____.

79) In $4 \times [6 + 3 \times (2 + 5)]$, the $6 + 3 \times (2 + 5)$ is enclosed in _____.

80) Dividing by two is equivalent to multiplying by one-_____.

81) Dividing by three is equivalent to multiplying by one-_____.

82) The fraction $\frac{1}{4}$ may be called one-_____ or a _____.

83) A value becomes _____ as large, or _____ what it was before, if it is multiplied by two.

84) 10^6 is a _____, 10^9 is a _____, and 10^{12} is a _____.

85) 0.01 is one _____ and 0.001 is one _____.

86) 10^{-6} is one _____ and 10^{-9} is one _____.

87) 12 is called a _____ and 144 is called a _____.

88) _____ is a fanciful name for 10^{100}, and _____ is a fanciful name for $10^{(10^{100})}$.

89) The _____ system uses base ten. The _____ system uses base two.

90) _____ numbers use base sixteen. _____ numbers use base eight.

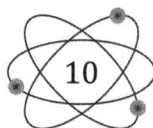

91) _____ refers to a system based on the number 60, such as dividing hours into minutes and dividing minutes into seconds.

92) $|-14|$ means to take the _____ of -14. (2 words)

93) In a^c, a is called the _____.

94) In a^c, a is _____ to the _____ of c.

95) In a^c, another name for c (aside from the answer to #94) is an _____.

96) $\frac{1}{5}$ is the _____ of 5. It could also be called a _____. (2 words)

97) Changing units from cm to in. or from cc to L (for example) is called a _____.

98) A _____ is a fraction made by dividing quantities that have different units, like distance and time or like weight and area.

99) _____ is a directed distance indicating the change in an object's position.

100) _____ is the instantaneous rate at which an object's position changes with respect to time, indicating how fast the object travels.

101) If the #100 is constant, it is equal to _____ over _____.

102) _____ is a combination of _____ and direction.

103) _____ is the instantaneous rate at which the first answer to #102 changes with respect to time.

104) The word _____ is used to describe a quantity that remains constant.

105) A _____ is either equal to a zero or a one.

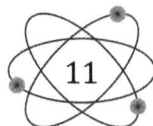

106) A _____ is equal to eight times the answer to #105.

107) _____ means to display information as a numerical value. (One alternative is #108.)

108) _____ means to display information using a dial (instead of #107).

109) To add fractions like $\frac{7}{8} + \frac{9}{16}$, you first need to find a _____. (2 words)

110) The diagram to the right for 30 is called a _____. (2 words)

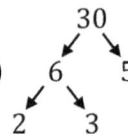

$$30 \nearrow \searrow$$
$$6 \quad 5$$
$$\swarrow \searrow$$
$$2 \quad 3$$

111) 8 is the _____ of 24 and 32. (3 words)

112) The _____ of 18 is $18 = 2 \times 3 \times 3$. (Just 1 word)

113) 12 is evenly _____ by 1, 2, 3, 4, 6, and 12.

114) 4, 9, 16, 25, 36, 49, 64, 81, 100, ... are _____. (2 words)

115) Rewriting $\frac{1}{\sqrt{2}}$ as $\frac{1}{\sqrt{2}} = \frac{1}{\sqrt{2}}\frac{\sqrt{2}}{\sqrt{2}} = \frac{\sqrt{2}}{2}$ is called _____ the denominator.

116) The _____ of an item is a quantitative measure of its worth.

117) Profit equals _____ minus _____.

118) The amount invested is called the _____. The _____ is how much money is earned (for an investment) or the cost of borrowing (for a loan).

119) _____ (2 words) and _____ (2 words) are two basic types of the second answer to #118.

120) A reduction in price is called a _____.

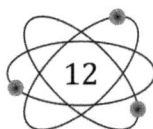

121) 3, 4, 5 and 5, 12, 13 are two examples of _____. (2 words)

122) A _____ number is equal to the sum of its divisors (excluding itself). For example, the divisors of 28 have a sum equal to $1 + 2 + 4 + 7 + 14 = 28$.

123) If the number described in #122 is even, it is called a(n) _____ number.

124) A _____ number can be expressed in the form $\frac{1}{2}(n)(n+1)$, such as 1, 3, 6, 10, 15, 21, 28, 36, and 45.

125) _____ numbers are a pair of numbers where each number is equal to the sum of the other number's divisors (excluding itself). One example is 220 and 284.

126) 29 and 31 as well as 41 and 43 are examples of _____. (2 words)

127) A pair of numbers are _____ if their greatest common factor is one. One example is 9 and 10. Another example is 8 and 15. (2 words)

128) $2^2 - 1 = 3$, $2^3 - 1 = 7$, and $2^5 - 1 = 31$ are examples of _____, but $2^4 - 1 = 15$ is not. (2 words)

6	7	2
1	5	9
8	3	4

1	2	3	4
2	1	4	3
3	4	1	2
4	3	2	1

A1	B2	C3	D4
B3	A4	D1	C2
C4	D3	A2	B1
D2	C1	B4	A3

129) The array of numbers on the left is called a _____ square.

130) The array of numbers in the middle is called a _____ square.

131) The array of numbers on the right is called a _____-_____ square.

132) Floating-point notation uses the form $a \times 10^n$, where a is called the _____ and is restricted by $0.1 \le a < 1$.

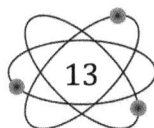

$$\begin{pmatrix} 1 & 4 & 3 \\ 2 & 0 & 8 \\ 5 & 6 & 7 \end{pmatrix} \qquad \begin{vmatrix} 1 & 4 & 3 \\ 2 & 0 & 8 \\ 5 & 6 & 7 \end{vmatrix}$$

133) The array on the left is a _____. The form on the right is a _____.

134) An _____ is a step-by-step procedure with specific instructions for solving a problem, often to be carried out using a computer program.

135) An _____ is the repetition of one or more steps in a procedure. Numerical methods that involve successive approximations typically feature this.

136) A typical _____ consists of a bar with a fixed point called a _____. Torques cause the bar to rotate (or pivot) about the fixed point.

137) _____'s method is an approximation method for finding the square root of a number based on the formula $x_{i+1} = \frac{1}{2}\left(x_i + \frac{n}{x_i}\right)$ for $\sqrt{n}$.

138) $a \times b = b \times a$ is the _____ property of multiplication.

139) $a + (b + c) = (a + b) + c$ is the _____ property of addition.

140) $a \times (b + c) = a \times b + a \times c$ is the _____ property.

141) $a + (-a) = 0$ and $a \times \frac{1}{a} = 1$ illustrate the _____ property.

142) If $a = b$ and $b = c$, then $a = c$ according to the _____ property.

143) $a = a$ is the _____ property.

144) A _____ is a chart or diagram that displays information.

145) One might _____ points, a curve, or a surface on the answer to #144.

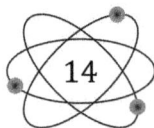

146) A _____ diagram shows relationships between sets with overlapping circles.

147) A _____ is a connected graph that doesn't have any cycles.

148) A _____ joins shapes (like triangles or hexagons) together at their edges to cover the plane, often in a repeated pattern.

149) A _____ chart divides a circle into sectors sized according to percentages.

150) A _____ chart uses rectangles to display a distribution of frequencies.

151) A _____ divides data into intervals (or bins) and uses rectangles to display a distribution of frequencies.

152) A _____ (2 words) consists of one rectangle and two _____. The rectangle represents the central half of the data. The second answer to this question helps to show how the central data relates to the minimum and maximum values.

153) A _____-and-_____ plot groups data according to place values. For example, for $35, 36, 38, 40, 41, 42$, and 43, the $35, 36$, and 38 would be grouped and would just show the $5, 6$, and 8 beside the 3 (showing that these are all in the thirties).

154) A _____ plot shows the relationship between (x, y) values. It is commonly used in experimental analysis for data points that approximately lie on a line or a curve.

155) The x-_____ is _____ and the y-_____ is _____ in the answer to #154. (The answers describe how these features are oriented.)

156) The features described in #155 include a _____, which includes tick marks and numbers to help read x- and y-values.

157) Each pair of (x, y) _____ forms a _____ on the graph.

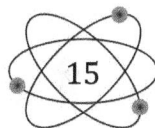

158) The symbols x and y are called _____. (2 words)

159) The values of (x, y) form an _____ pair.

160) The point $(0, 0)$ is called the _____.

161) The x-_____ is called the _____.

162) The y-_____ is called the _____.

163) The x- and y-_____ divide the xy plane into four _____.

164) A _____ plot includes one standard scale and one logarithmic scale.

165) The y-_____ is the value of y for which a line or curve crosses the y-axis.

166) A graph of $ax + by + c = 0$ in the xy plane looks like a _____ (2 words).

167) The answer to #166 has constant _____ (or steepness or gradient).

168) The answer to #167 equals the _____ over the _____.

169) To _____ means to estimate a value that lies between tick marks.

170) To _____ means to estimate a value that lies beyond the range shown.

171) A _____ is a locus of points in a plane equidistant from a common center.

172) In #171, a _____ extends from the center to a point on the curve.

173) The _____ is twice as long as the answer to #172.

174) A _____ is a line that touches the answer to #171 at a single point.

175) A _____ is a line that crosses the answer to #171 at two points.

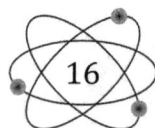

176) A _____ is a line segment that joins two points on the answer to #171.

177) The path along a curve between two points is called an _____.

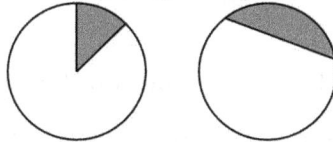

178) The shaded region on the left is a _____ and on the right is a _____.

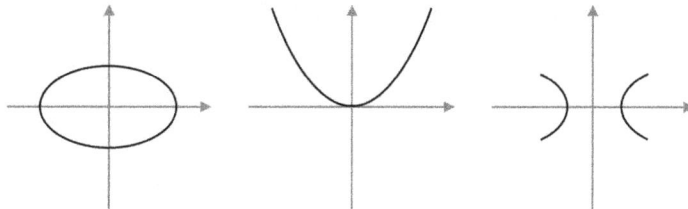

179) The curve on the left resembles a(n) _____.

180) The curve in the middle resembles a(n) _____.

181) The curve on the right resembles a(n) _____.

182) The curves on the left and right each have two _____.

183) For the left curve, the long axis is called the _____ axis.

184) The two separate parts of the right curve are called _____.

185) The middle curve features a _____ (not shown in the diagram), which is a mathematically significant horizontal line.

186) The three curves are distinguished by how their _____ compares to unity.

187) The curve on the right features two slanted _____ (not shown).

188) The three curves shown previously are _____ (2 words), which can be obtained as the intersection of a plane and a double _____.

189) A closed graph with at least one edge is called a _____.

190) An _____ is traced out by a point on the edge of one circle that rolls around the edge of another circle (which may have different size).

191) Ancient Greeks used the answer to #190 and another circle called the _____ in order explain the apparent motion of the sun and planets in their geocentric theory.

192) A line of _____ divides a figure into two mirror images.

193) #192 is also referred to as _____. (2 words)

194) A _____ is a connected subset of 2D space, such as the interior of a circle.

195) A curve is said to be _____ if the curve is continuous, if it lies in the plane, and if it doesn't intersect itself.

196) A _____ refers to a curve that loops itself (without intersecting itself) such that when the two ends join together, the curve can't be untangled into a simple loop.

197) _____ serves as a simple model for population increase. (2 words)

198) _____ is indicative of radioactive nuclear decays. (2 words)

199) A _____ involves an object moving one step (of fixed size) at a time, where pure chance is used to select the direction that the object moves each step. (2 words)

200) Two curves are said to _____ if the curves have the same tangent at the point where they meet.

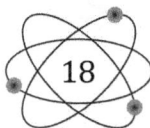

201) _____ refers to the measurements (and calculations that relate to these) of angle, length, area, and volume of various figures.

202) _____ features "parallel" lines that appear to meet at a _____ point. It is the art of drawing realistic diagrams of 3D objects on a 2D surface.

203) A straight _____ continues infinitely in both directions. A _____ (2 words) is the finite portion of the first answer to this question that connects two points.

204) A _____ is straight, but semi-infinite. It has an endpoint at one side, while extending infinitely in the other direction.

205) Two lines are _____ if they are always the same distance apart.

206) Two lines _____ if they cross at one point.

207) Two lines are _____ if they are neither the answer to #205 nor #206. (This is only possible in 3D space. It's not possible if both lines lie in the same plane.)

208) Two lines are _____ if they meet at a right angle.

209) Other words to describe #208 include _____ and _____.

210) The point where the curves meet in the right figure is called a _____.

211) Two lines that extend from the same point form an _____.

212) The point where two lines cross is called a _____ or _____.

213) Each line in #211 is called an _____.

214) One unit for measuring #211 is found by dividing a circle into 360 _____.

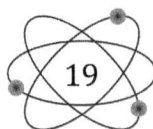

215) One sixtieth of #214 is called an _____. (2 words)

216) One sixtieth of #215 is called an _____. (2 words)

217) The main alternative to #214 is called a _____.

218) One hundredth of a right angle is known as a _____.

219) A ship's _____ is the direction that its course makes clockwise from north.

220) A _____ is a 2D figure bounded by line segments.

221) The line segments (or edges) that bound the answer to #220 are called _____.

222) A _____, _____, _____, and _____ are examples of #220 bounded by three, four, five, or six (not necessarily equal) line segments.

223) A _____, _____, _____, and _____ are examples of #220 bounded by seven, eight, nine, or ten line segments.

224) A _____ and _____ are examples of #220 bounded by eleven or twelve line segments.

225) A _____ has four equal sides and right angles. A _____ has four right angles and two pairs of equal sides.

226) A _____ has two pairs of parallel edges. A _____ is a special case of this with four equal sides.

227) A _____ has one pair of parallel edges and four sides in total.

228) A _____ has two pairs of adjacent sides with equal length.

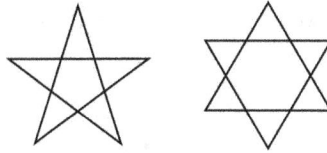

229) The left figure (which doesn't have the word "star" in it) is called a _____.

230) The right figure (which doesn't have the word "star" in it) is called a _____.

231) A figure bounded by three edges is _____ if it includes a 90° angle.

232) A figure bounded by three edges is _____ if each angle is less than 90°.

233) A figure bounded by three edges is _____ if one angle is greater than 90°.

234) A _____ is greater than 180° and less than 360°. (2 words)

235) The answer to #234 is called re-_____ if it is an interior angle of a 2D figure bounded by straight edges.

236) The longest side of the answer to #231 is called the _____.

237) The short sides of the answer to #231 are called _____.

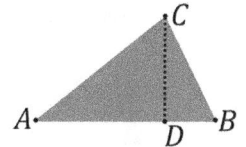

238) *AB* is the _____ and *CD* is the _____ or _____ of the triangle shown to the right. We could use these distances to determine the area.

239) Point *C* is called the _____ for the orientation of the triangle shown above.

240) _____ means to cut a figure into two equal parts.

241) _____ means to cut a figure into three equal parts.

242) The _____ of a line segment lies at the middle of the line segment.

243) A _____ joins the corner of a three-sided polygon to the point on the middle of the opposite side.

244) A _____ connects two opposite corners of a rectangle.

245) _____ is the total distance around the edges of a bounded plane figure.

246) _____ provides a measure of the 2D space bounded by a closed plane figure or the 2D space on the surface of a 3D object.

247) _____ provides a measure of the 3D space enclosed by a 3D object.

248) Two _____ angles form a right angle. Two _____ angles form a straight line (or a "straight angle").

249) _____ angles form on opposite sides of a point where lines intersect.

250) An _____ angle forms at a corner of a polygon and lies inside the polygon.

251) An _____ angle forms between an edge of a polygon and a line that extends from an adjacent edge.

252) A _____ is a flat 2D space, such as the one containing the x- and y-axes.

253) A _____ rectangle is a special rectangle in that if a square is removed from its shorter end, the remaining rectangle has the same shape as the original rectangle.

254) If you cut a circle in half, you get two _____.

255) The total distance around the edge of a circle is called the _____.

256) The ratio of the answer to #255 to the diameter of a circle is called _____.

257) The distance along a section of a curve is called _____. (2 words)

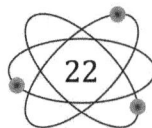

258) The angle between two radii of a circle is called a _____ angle.

259) An _____ angle forms between two chords of a circle that share a common endpoint.

260) Each side of a three-sided polygon _____ the angle opposite to it. The same answer may also be used to describe the angle formed by the arc of a circle, for example.

261) An _____ polygon has edges with the same length.

262) An _____ polygon has interior angles with the same angular measure.

263) A _____ polygon satisfies both #261 and #262.

264) An _____ three-sided polygon has two edges with the same length.

265) A _____ three-sided polygon has edges with three different lengths.

266) _____ means the same distance from two or more points (or figures).

267) A _____ is a set of points in space that satisfy one or more conditions.

268) _____ and _____ describe the two different ways that an object can travel along the edge of a circle.

269) The region lying between two concentric circles is called the _____.

270) A closed 3D figure (whether curved or flat) is referred to as a _____.

271) The 2D region surrounding the answer to #270 is called the _____.

272) A figure bounded by six perpendicular square sides is called a _____.

273) A rectangular box is called a _____.

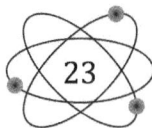

274) One kind of _____ is a 3D object with a square base and triangular sides.

275) One kind of _____ is a 3D object with triangles on opposite sides joined together by three rectangles.

276) A _____ is a 3D object with 6 octagons and 8 triangles. (2 words)

277) A _____ is the locus of points (in 3D) equidistant from a common center.

278) If you cut the answer to #277 in half, you get two _____.

279) The locus of points contained inside the answer to #277 is called a _____.

280) The shape is called a _____ if there is a hollow region in the center of the answer to #279. (2 words)

281) Circles of _____ are horizontal circles covering earth's surface.

282) Circles of _____ are vertical circles covering earth's surface.

283) The circles of #282 intersect at two points called _____.

284) A _____ is a circle on the surface of a sphere which shares the same center as the sphere. (2 words)

285) When applied to the earth, the answer to #284 is called a _____.

286) The _____ is the special circle of #281 that is also #284.

287) _____ is a method for determining the distance to an unknown location by using angles measured from two known locations.

288) A simple _____ can be formed by curving a rectangular sheet such that one edge joins an opposite edge, and fastening two circles to the ends.

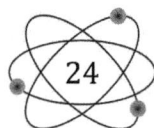

289) A simple _____ joins a circular base to a point directly above the center of the base by connecting straight line segments from the point to the base.

290) The length of any of the line segments in #289 is a _____. (2 words)

291) You get a _____ if you slice #289 into two pieces with a plane parallel to the base (and discard the top portion).

292) A _____ is a shape that can be formed by rolling or bending a plane surface without cutting or stretching. Examples include the answers to #288 and #289, but not the answer to #277. (2 words)

293) A _____ is a shape that can be traced out by moving (such as rotating or translating) a line segment. Examples include the answers to #288 and #289. (2 words)

294) The locus of points lying inside of a circle is called a _____.

295) The shape is called a _____ if a smaller circle is cut out of the center of the answer to #294.

296) If it is very thin, #295 is sometimes referred to as a _____.

297) A _____ is a 3D object with the shape of a doughnut.

298) A _____ is a curve that increases in size as it winds around.

299) A _____ is a curve that winds around the answer to #288 at a constant angle.

300) Points (or line segments) are _____ if they lie on the same line.

301) Figures are _____ if they lie in the same plane.

302) Figures are _____ if they occupy the same points in space.

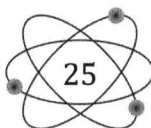

303) Two or more lines are _____ if they all pass through the same point.

304) Circles (or spheres) are _____ if they share a common center.

305) Cylinders are _____ if they share a common axis.

306) The balancing point of an object is the object's _____. (3 words)

307) For a triangle, the answer to #306 lies at the _____.

308) A _____ is the path of shortest distance constrained to lie on a surface.

309) The xy, yz, and zx planes divide 3D space into eight _____.

310) A _____ is a 3D figure that is bounded by flat sides.

311) The sides of the answer to #310 are called _____.

312) The five regular convex kinds of #310 are called _____. (2 words)

313) A set of 13 semi-regular kinds of #310 are called _____. (2 words)

314) The type of #312 bounded by 4 triangles is the _____.

315) The type of #312 bounded by 8 triangles is the _____.

316) The type of #312 bounded by 12 pentagons is the _____.

317) The type of #312 bounded by 20 triangles is the _____.

318) The type of #313 bounded by 12 pentagons and 20 triangles is the _____.

319) The type of #313 bounded by 4 hexagons and 4 triangles is the _____. (2 words)

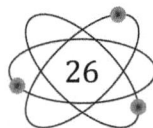

320) The type of #310 bounded by 6 rhombuses is the _____.

321) A _____ is a flat strip in the shape of a loop containing one twist so as to only have one surface and one edge. (2 words)

322) A _____ is a curved surface with one surface and no edges. (2 words)

323) _____ means rectangular.

324) An _____ is a line segment that connects the centroid of a regular polygon to the center of one of its sides.

325) A _____ refers to a side of a 3D solid that isn't its base. (2 words)

326) A _____ of an object (or of a space) is a measure in which it extends. For example, a plane has two of these whereas a line has only one.

327) A _____ is a curve or surface with an infinite pattern of self-similarity. Its number of #326's is technically not an integer.

328) A _____ is a plane generalized to 4D space. An example of one is 3D space.

329) A _____ is a cube generalized to 4D. A _____ is an n-cube.

330) A _____ is a sphere generalized to 4D. A _____ is an n-sphere.

331) A _____ is an object bounded by flat sides in n-dimensional space.

332) A _____ is a tetrahedron generalized to n dimensions.

333) The _____ for a polyhedron is $\{E, F\}$, where E is the number of edges on each face and F is the number of faces meeting at each vertex. (2 words)

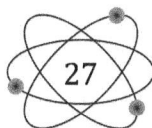

334) _____ is the branch of mathematics that involves the most basic operations (like addition and multiplication) and powers.

335) _____ is the branch of mathematics devoted to money management.

336) _____ is the study of wealth (in the form of goods and services), including its production, consumption, and distribution.

337) _____ is the branch of mathematics that represents arithmetic operations using letters for unknown quantities.

338) _____ is an advanced area of #337 involving groups, rings, and fields. (2 words)

339) _____ is an advanced area of #337 involving matrices. (2 words)

340) _____ is the branch of mathematics that studies points, lines, curves, and shapes, along with their properties.

341) _____ is the branch of mathematics that specializes in coding messages or deciphering coded messages.

342) _____ theory is the branch of mathematics that studies events which occur through random chance.

343) _____ is devoted to counting and arranging. It is useful for #342.

344) _____ is the branch of mathematics devoted to data collection and analysis.

345) _____ mathematics studies natural phenomena and real-world applications.

346) _____ is the branch of mathematics involving geometric figures and spaces, including which properties are preserved under various transformations.

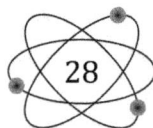

347) _____ theory is the branch of higher mathematics that deals with positive integers.

348) _____ theory is the branch of mathematics devoted to determining the best possible selection from available choices.

349) _____ theory is the branch of mathematics that involves determining which strategy will result in the best outcome. It has been applied to recreational activities, war, business, and other areas.

350) _____ is the technique used in problem-solving that is based on previous experience and trial-and-error. It helps to analyze complex data more efficiently.

351) _____ is the branch of mathematics stemming from the ratios of the sides of a right triangle.

352) _____ theory is the branch of mathematics where a small change in the initial conditions of a complex system can result in dramatically different outcomes.

353) _____ translates statements made in ordinary language to a concise mathematical form to develop rules of reasoning. (2 words)

354) _____ is the branch of mathematics that begins by assigning the value of one to "truth" and the value of zero to "false." (2 words)

355) _____ theory is the branch of mathematics devoted to collections of objects that are well-defined.

356) _____ theory is the branch of mathematics devoted to collections of objects that satisfy associativity, invertibility, identity, and closure.

357) _____ is the branch of mathematics involving derivatives and integrals.

358) _____ is the branch of mathematics devoted to solving problems that feature first and second derivatives. (2 words)

359) _____ theory is the branch of mathematics that studies networks, which are structures consisting of vertices (called nodes) and edges (called links).

360) _____ is the branch of mathematics specializing in the techniques of approximation, using computers to solve complex real-world problems. (2 words)

361) _____ applies statistics to the field of biology.

362) An _____ or _____ is a statement that is either deemed to be self-evident or is assumed to be true, from which other principles may be derived.

363) _____ are foundational assumptions which can't be deduced from other assumptions. (2 words)

364) A _____ is a mathematical statement that can be proven from #363's using logic and reason.

365) A _____ applies logic, reason, given information, #363's, and #364's to draw a conclusion about a mathematical statement.

366) A generalized or important #364 is sometimes called a _____ in math. (In science, this term would require extensive experimental confirmation.)

367) A _____ #365 supposes that p is true and uses this to show that q is true.

368) An _____ #365 supposes that q is false and uses this to show that p is false, in order to demonstrate that q follows from p.

369) A _____ #365 is deduced from axioms, whereas an _____ #365 is formed from algebraic relations and analysis only.

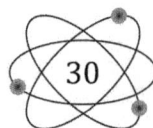

370) A _____ is a mathematical or logical statement that is to be proven true or false.

371) _____ is a mathematical statement that hasn't been proven. It may be based on a partial pattern or incomplete information. In a sense, it may seem synonymous with educated guesswork or speculation (but the term "hypothesis" is more common in science).

372) A _____ (also called a helping _____) is a mathematical statement that is mainly proven in order to help prove another more significant statement.

373) A _____ is a mathematical statement that follows from a proven statement with very little effort.

374) A result known to be true in a specific situation is said to be _____ once it has been reformulated in a way that it applies to a wide variety of situations.

375) _____ is the process of applying reason to a set of premises in order to draw a conclusion from them which is logically valid.

376) Mathematical _____ first demonstrates that a property holds for zero, and then proves that if the property holds for the number n, it must also hold for n + 1. In this way, the property is shown to hold in general.

377) Statistical _____ is a conclusion that is drawn based on a sample.

378) A _____ shows T' and F's for a compound statement. (2 words)

379) "This statement is false," is known as the _____. (2 words)

380) An error in reasoning is called a _____.

381) An _____ conclusion can be drawn from #380.

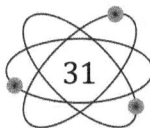

382) A _____ statement has the form, "If A is true, then B is true."

383) In #382, A is called the _____ and B is called the _____.

384) Given the statement, "If A is true, then B is true," the _____ is, "If B is true, then A is true." (This may be invalid even if the original statement is true.)

385) Given the statement, "If A is true, then B is true," the _____ is, "If A is false, then B is false." (This may be invalid even if the original statement is true.)

386) Given the statement, "If A is true, then B is true," the _____ is, "If B is false, then A is false." (This holds if the original statement is true.)

387) If X represents a statement, writing "not X" represents the act of _____.

388) The statement, "If A is true, then B is true," means that A _____ B.

389) The statement, "A _____ B," means both, "If A is true, then B is true," and, "If B is true, then A is true." (4 words)

390) A _____ condition must be true in order for a statement to be true.

391) The statement described in #389 is a _____ and _____ condition.

392) One example of a mathematical _____ is, "A triangle is to a hexagon as a quadrilateral is to an octagon."

393) _____ is knowledge obtained independent of experience. (2 words)

394) _____ is knowledge gained through experience. (2 words)

395) A _____ draws a conclusion by essentially using the same conclusion as part of the proof. (2 words)

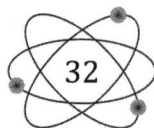

396) A theory is consistent provided that it doesn't include a _____.

397) A _____ occurs when an unreasonable conclusion is drawn from a set of reasonable assumptions.

398) A statement is termed _____ if the statement is not necessarily true or false.

399) A _____ is a special case that proves a statement false.

400) _____ is something which produces an effect.

401) A _____ statement is built from simpler statements with words like "and."

402) A _____ uses the word "and" to join two simpler statements together.

403) An _____ uses the word "or" to join two simpler statements together such that it is true if at least one of the simpler statements is true. (2 words)

404) An _____ involves two simpler statements and is true if exactly one of the simpler statements is true (but not both). (2 words)

405) Two events are _____ if only one event or the other can occur, but not both. (2 words)

406) Two #401's are said to be _____ if they have the same components and also have identical truth tables. (2 words)

407) Two statements are said to be _____ if either statement can be deduced from the other (that is, they are logically equivalent in meaning).

408) A _____ is a #401 that is true for all possible truth values of its components.

409) This sentence is an example of _____. (hyphenated)

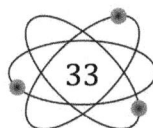

410) A _____ draws a conclusion that follows from two premises. For example, "All birds have wings. All owls are birds. Therefore, all owls have wings."

411) A _____ is the division of a whole into two parts.

412) _____ is Latin for "which was to be proved." (3 words)

413) _____ is Latin for "which was to be done." (3 words)

414) _____ is Latin for "reduction to absurdity." (3 words)

415) _____ could mean a hundredth or it could mean divided into hundredths.

416) A _____ number is the minimum number of different colors needed such that the vertices can be colored in a way that no two adjacent vertices have the same color.

417) _____ motion is movement along a straight line (forward or backward).

418) A _____ (or an _____) is represented by a symbol like x or y.

419) If a symbol has a fixed numerical value, it is called a _____.

420) A number that multiplies a variable, like the 3 in $3x$, is called a _____.

421) When exploring how a affects the graph of $y = ax$, a is called a _____. When graphing $x = \cos t$ and $y = \sin t$, the symbol t would also be called this term.

422) $3x - 4 = 8$ is an _____ whereas $5x + 4$ is an _____.

423) The _____ (of the second answer to #422) are separated by $+$ or $-$ signs.

424) $2x < 9$ is an _____ and $3 < x < 7$ establishes an _____.

425) Information that is stated in a problem is said to be _____.

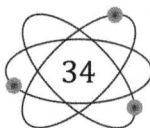

426) You can _____ $4x - 5 = 15$ for x. You can _____ $8x - 3x + x$.

427) _____ is the phrase used to rewrite $5x + 9 + 3x - 8 - 2x + 3$ as $6x + 4$. (3 words)

428) _____ is the phrase used to describe how $5x - 4 = 3x + 12$ is rewritten as $2x = 16$ and then $x = 8$. (3 words)

429) $x = 3$ and $x = -1$ are two _____ to $x^2 - 2x - 3 = 0$.

430) $x = y = 0$ is the _____ to $3x - 2y = 0$. (2 words)

431) In the case of $0x = 1$, we say that x is _____.

432) In the case of $0x = 0$, we say that x is _____.

433) $x + 1 = x + 2$ has _____. (2 words)

434) _____ satisfy $x + x = 2x$. (3 words)

435) $x^2 = -1$ is said to be _____ with respect to the set of real numbers.

436) $x = y + 2$ and $x = y + 1$ are said to be _____ or _____ since if you subtract the equations you get $0 = 1$, which is impossible.

437) The system $x + y + z = 1$ and $x - y = 2$ is _____ because there are not enough equations to solve the problem. (hyphenated)

438) $e^x + x = 0$ is an example of a _____ equation because you can't isolate x by applying algebra. You need numerical techniques to determine x.

439) Practical equations like $P = 2L + 2W$ and $H = \frac{1}{2}gt^2$ are called _____.

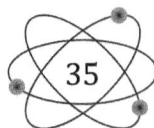

440) _____ means to determine the numerical value of.

441) _____ refers to exercises that only involve inserting numbers in place of symbols and then calculating the answer. (3 words)

442) We _____ the $2x$ when we write $2x(x + 3) = 2x^2 + 6x$.

443) We _____ the x when we write $x^2 + 4x = x(x + 4)$. (2 words)

444) $x^2 + 2x + 1$ is the _____ form of $(x + 1)^2$.

445) $ax + b = 0$ has the form of a _____ equation, provided that $a \neq 0$.

446) $7x^8 + 5x^6 + 3x^4$ and $x^3 - 2x + 4$ are examples of _____.

447) $ax^2 + bx + c = 0$ is a _____ equation, provided that $a \neq 0$.

448) For #447, $b^2 - 4ac$ is called the _____.

449) The _____ of a #446 refers to the highest power of any of its terms.

450) A _____, _____, and _____ are #446's to the third, fourth, and fifth #449, respectively.

451) $x - 1 - x$ and $\frac{6x}{2x}$ are examples where x _____ out.

452) $x^2 - 1$ and $x^2 + 6x + 9$ are said to be _____ because they can be rewritten in the forms $(x + 1)(x - 1)$ and $(x + 3)(x + 3)$, respectively, in contrast to $x^2 + 5$.

453) We _____ to rewrite $\frac{2}{x} = \frac{y}{3}$ in the form $2(3) = xy$. (hyphenated)

454) The system $2x + y = 8$ and $x - y = 1$ are called _____ (if we are to find the values of x and y which satisfy both equations).

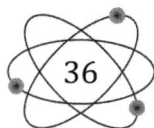

36

455) We say that y has been _____ from the equations when we arrive at $19x = 38$ (after doing some algebra), having begun with $4x + 3y = 11$ and $5x - y = 9$.

456) $2x + y = 9$ and $x - y = 3$ is solved by the method of _____ if we write plug $y = 9 - 2x$ into $x - y = 3$ to get $x - (9 - 2x) = 3$ and proceed to find x and y.

457) It is called _____ the _____ when we rewrite $x^2 + 8x$ as follows: $x^2 + 8x + 16 - 16 = (x + 4)(x + 4) - 16 = (x + 4)^2 - 16$.

458) $x + y$, $x - 2$, $x^4 + x^2$, and $x^2 - y^2$ are examples of _____. In contrast to these, $6x + 2x$, $x + y + z$, and $x^3 + x^2 + x$ are not.

459) _____ is formed by the coefficients of $(x + y)^2$, $(x + y)^3$, $(x + y)^4$, $(x + y)^5$, and so on when they are expanded out. (2 words)

460) $x + y + z$ and $x^2 + x + 1$ are _____ (which are analogous to #458).

461) x and y are _____ related if an increase in one causes the other to increase.

462) x and y are _____ in $y = ax$. (2 words)

463) x and y are _____ in $y = \frac{a}{x}$. (2 words)

464) Algebraic _____ include $x^2 - y^2 = (x + y)(x - y)$ and $(a + b)(c + d) = ac + ad + bc + bd$, for example. Such formulas hold for all values of the symbols.

465) _____ is a method (using a table) to solve a problem of the following form: $(x^3 + x^2 + x - 3) \div (x - 1)$. (2 words)

466) $x^4 + 3x^3y + 6x^2y^2 + 3xy^3 + y^4$ is _____ in the _____ degree.

467) $2 < x < 3$ is an _____ whereas $2 \le x \le 3$ is a _____. (2 words each)

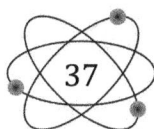

468) An _____ is true for any real value of the variable. For example, any real value of x makes $x^2 > -1$ true, whereas $x > -1$ only holds for some values. (2 words)

469) The _____ form of an equation (or expression) is its standard form.

470) $x + \sqrt{a}$ and $x - \sqrt{a}$ are known as _____. (2 words)

471) When there is exactly one solution to a problem, it is said to be _____.

472) Two answers to a problem are _____ if their numerical values differ.

473) When two (or more) roots turn out to be identical, they are called _____.

474) A problem includes _____ (or _____) information if some of the given information isn't needed to determine the answer.

475) The equations $4x + 2y = 14$ and $8x + 4y = 28$ are said to be _____ since the second equation doesn't provide anything new compared to the first equation.

476) _____ equations refer to a set of equations where one of the equations is a linear combination of the other equations.

477) A _____ equation has integer coefficients and integer solutions.

478) It is _____ to solve $x^2 = 4$, but not $x^2 = -4$ (using real numbers).

479) _____ figures have the same size and shape.

480) _____ figures have the same shape, but not the same size.

481) For a _____ polygon, every interior angle is less than $180°$.

482) For a _____ polygon, at least one interior angle is greater than $180°$.

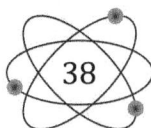

483) The word _____, used to describe some polyhedra, means star-shaped.

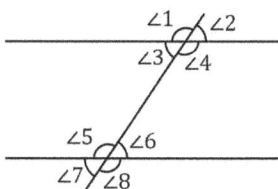

484) In the diagram above, the _____ is the line that intersects two other lines.

485) ∠2 and ∠6 above (where two lines are parallel) are called _____ angles.

486) ∠3 and ∠6 above are called _____ angles. (2 words)

487) ∠2 and ∠7 above are called _____ angles. (2 words)

488) A polygon is _____ in a circle (or the circle is _____ about the polygon) when all of the polygon's vertices lie on the arc of the circle.

489) A _____ curve begins and ends at the same point, enclosing an area.

490) A _____ is a polyhedron bounded by 6 parallelograms (not necessarily right).

491) A _____ is a polyhedron bounded by 6 squares and 8 triangles.

492) In geometry, to _____ a figure means to draw it using a straight edge and compass without measuring distances or angles.

493) _____ is an angle measured upward relative to a horizontal line.

494) The angle of _____ is measured downward relative to a horizontal line.

495) Two _____ angles add up to 360°.

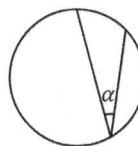

496) In the diagram to the right, α is an _____ angle.

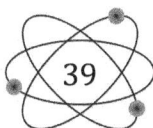

497) Two ellipses or hyperbolas are _____ if they share the same foci.

498) A collection of points are _____ if a circle passes through all of them.

499) _____ refers to the intersection (or partial overlap) of two figures.

500) Two (or three) lines or planes are said to be _____ if one is slanted where they intersect (that is, if they aren't mutually perpendicular).

501) A _____ angle is formed by two half planes.

502) A simple definition of _____ is the number of holes that a surface has.

503) The _____ point of a triangle is the point that minimizes the total distance from the point to all three vertices.

504) The shadow that a 3D object casts on a plane surface is called its _____.

505) The figure of intersection of a plane and a 3D object is called a _____.

506) On a sphere, the _____ is the region between two parallel planes.

507) The _____ axis of two circles is the locus of points for which the lengths of the tangents to the two circles are equal.

508) The words _____ and _____ are sometimes used to describe the chance of an event producing a given outcome (even though their precise meanings differ).

509) In the context of betting, #508 is called the _____.

510) #508 is described as _____ when it is based on observations.

511) The outcome of rolling a fair die is _____ if each result is equally likely.

512) A _____ is the arrangement of elements in a particular order.

513) A _____ is like #512, except that order doesn't matter.

514) Observations made during an experiment are called _____.

515) _____ refers to using equipment to gather a quantitative form of #514.

516) #514 is termed _____ if it hasn't been processed, analyzed, screened, etc.

517) _____ means to transform #514 into coded form (as a measure of security).

518) A _____ displays #514 in rows and columns.

519) The _____ refers to the entire set of values. A _____ is a portion of the full set.

520) The second answer to #519 is _____ of the first answer to #519 if it shares the same characteristics.

521) A single observation (or test) is referred to as a _____.

522) When repeating an experiment, if the original results are _____, this helps to confirm our understanding.

523) A statistical _____ or _____ is an extreme #514, lying outside of the main body. (These terms are sometimes used interchangeably, yet differ in meaning.)

524) The _____ is a simple average obtained by adding up all of the values and dividing by the number of values. (2 words)

525) The _____ (of positive values) is found by multiplying the values and then taking the nth root (where n is the number of values) (2 words)

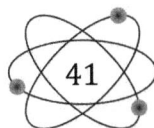

526) The _____ takes the reciprocal of the given values, finds the #524 of these reciprocals, and then takes the reciprocal of the result. (2 words)

527) The _____ multiplies each given value by a factor (in the sum in the numerator) and divides by the sum of these factors. (2 words)

528) When the values are arranged in order, the _____ is the middle value if the number of values is odd, or the average of the two middle values if it is even.

529) The _____ is the number of times a particular data value is observed.

530) The _____ equals a particular #529 over the sum of #529's. (2 words)

531) The _____ is the result for which #529 is the highest.

532) A _____ distribution of #529's has two peaks.

533) #514 is said to be _____ (or _____) if it is qualitative/descriptive.

534) The _____ is an incorrect notion that the outcome of an event has a higher probability of occurring if it has been under-observed thus far. (3 words)

535) The _____ variable is manipulated and the _____ variable changes in response during an experiment.

536) The _____ is the difference between exact and approximate values.

537) _____ #536's cause fluctuations about the true value.

538) _____ #536's affect the results in a consistent direction.

539) A graph of the fluctuations in #537 would be described as _____.

540) If you divide #536 by the exact value, you get the _____ #536.

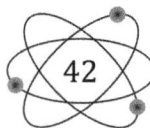

541) _____ #536 multiplies #540 by 100%.

542) The _____ is the difference between an observed value and predicted value in statistics.

543) Ranked values are sometimes divided into four equal parts called _____.

544) A _____ is like #543, but divided into one hundred equal parts.

545) The _____ is the difference between the maximum and minimum values.

546) The _____ is the difference between the upper and lower #543's. (2 words)

547) The _____ is the difference between a particular observation and the mean.

548) The value predicted for a random variable is called the _____. This is also known as the mean. (2 words)

549) The #548 of the square of the #547 is called the _____.

550) The square root of #549 is called the _____. (2 words)

551) #545, #546, and #550 provide measures of the _____ or _____.

552) _____ statistics are largely unaffected by #523's.

553) _____ refers to experimenting with a variety of methods until one of the methods works well. (3 words)

554) The goodness of _____ indicates how well a model fits the data.

555) The method of _____ minimizes the sum of the squared differences between the observed and predicted values. (2 words)

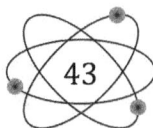

556) A _____ can be used to determine how well data fits a straight line. (2 words)

557) A _____ test helps to assess #554. (hyphenated)

558) _____ indicates the likelihood that the true value lies in an interval. (2 words)

559) _____ indicates the likelihood that similar results would be produced under the same conditions.

560) The _____ between random variables indicates how a change in one would cause a change in the other.

561) #560 is described as _____ when the link occurs through another variable.

562) A _____ variable is a factor not under investigation during an experiment which may affect the outcome of the experiment.

563) #562 is described as _____ when it is unknown and unaccounted for.

564) The _____ group provides a baseline for making comparisons.

565) _____ indicates the level of asymmetry in the distribution of data.

566) Lack of objectivity in data analysis is referred to as _____.

567) _____ indicates how well a measurement agrees with the true value.

568) _____ indicates how well multiple measurements agree with each other.

569) The _____ refers to the number of independent quantities affecting a statistical calculation. (3 words)

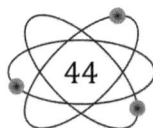

570) The _____ (also called _____) is a bell-shaped curve that is used widely in statistics. (2 words each)

571) A variable is said to have been _____ if it has been transformed such that it now has a mean equal to zero and a variance equal to unity.

572) A mathematical _____ is an abstract problem that approximates a problem in the real world.

573) A computer _____ aims to mimic behavior in the real world.

574) A _____ process is determined randomly.

575) A well-defined collection of distinct objects is called a _____.

576) A _____ is a well-defined collection of distinct objects that satisfies associativity, invertibility, identity, and closure.

577) An object in a #575 is called an _____ or a _____.

578) The _____ of two 575's combines all of the objects from each #575.

579) The _____ of two 575's only includes objects that belong to both 575's.

580) Zero objects belong to the _____ (or _____) #575.

581) The _____ (or _____) doesn't change an object that is combined with it in a binary operation. (2 words each)

582) The _____ effectively "undoes" the effect involved in a binary operation. (2 words)

583) _____ results in a mirror image.

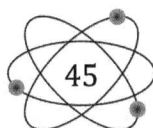

584) _____ moves an object along a straight line.

585) _____ about an axis is circular motion.

586) A _____ transforms a rectangle into a (non-right) parallelogram.

587) An _____ relation is a binary operation that is reflexive, symmetric, and transitive.

588) The _____ number is the number of distinct objects in a #575.

589) Two #575's are said to be _____ if they don't have any objects in common.

590) Objects a, b, and c have _____ order if abc, bca, and cab are the same, and are opposite to acb, bac, and cba.

591) A #575 is called a _____ if each object in it is itself a #575.

592) A #575 is basically _____ if it can be "counted." (This is informal.)

593) The term _____ relates the objects of two different #575's.

594) ab, bc, and ca are the _____ products of the #575 {a, b, c}.

595) The _____ #575 is the #575 of objects that are in one set, but not another.

596) _____ uses real values from 0 to 1 (instead of just the digits 0 and 1) for false and true, allowing for "partial truth." (2 words)

597) A #575 that contains exactly one object is called a _____.

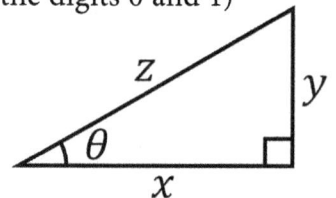

598) x is the _____, y is the _____, and z is the _____ in the diagram above (relative to θ).

599) For the previous diagram, the ratio of y to z is called _____.

600) For the previous diagram, the ratio of x to z is called _____.

601) For the previous diagram, the ratio of y to x is called _____.

602) For the previous diagram, the ratio of z to y is called _____.

603) For the previous diagram, the ratio of z to x is called _____.

604) For the previous diagram, the ratio of x to y is called _____.

605) The _____ is the inverse of the function in #599.

606) There are generally two answers for the function in #605. Of these, the _____ value is the one that would be returned by a typical calculator.

607) A graph that resembles that of the function in #599 is said to be _____, even if it shifted. (The term also applies if it is technically #600.)

608) The _____ is centered at the origin and has a radius of one. (2 words)

609) One example of _____ equations is $x = a \cos \theta$ and $y = a \sin \theta$.

610) A _____ includes both a magnitude and direction. A _____ has a magnitude, not direction. (The technical distinction involves a coordinate transformation.)

611) Two <the first answer to #610>'s that are opposite are said to be _____.

612) The _____ of a <the first answer to #610> can be found by projecting it onto the x-, y- and z-axes.

613) A _____ is a <the first answer to #610> that has a magnitude of one unit. (2 words)

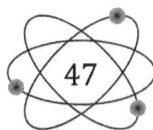

614) The _____ is the sum of the absolute values of #612's. (2 words)

615) We say that a <the first answer to #610> is _____ into its #612.

616) The _____ is the combination of two or more vectors joined tip-to-tail.

617) The _____ is a straight line from the initial position to the final position.

618) _____is the maximum displacement from equilibrium.

619) A simple #599 wave goes through one complete _____ every 2π radians.

620) #619 is called a _____ in the context of circular motion.

621) _____ is the time it takes to complete one oscillation.

622) _____ is the number of cycles per unit time.

623) _____ is the distance that a wave travels in one cycle.

624) _____ is basically #622 converted from cycles/sec to rad/sec. (2 words)

625) The _____ angle shifts a #607 wave horizontally.

626) An _____ is a back-and-forth motion.

627) _____ motion can be described with #599 and #600 functions.

628) Motion is _____ if it is repeated in equal time intervals.

629) A _____ expresses a relationship between one variable and another.

630) The possible input values for #629 are called the _____.

631) The possible output values for #629 are called the _____.

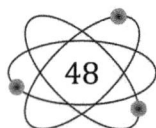

632) If $f(-x) = f(x)$ for all x (in its #630), then $f(x)$ is an _____ #629.

633) If $f(-x) = -f(x)$ for all x (in its #630), then $f(x)$ is an _____ #629.

634) The _____ of a graph in a particular neighborhood indicates if the slope of a #629 is strictly increasing or decreasing in that neighborhood.

635) The #634 may be described as _____ or _____. (2 words each)

636) $f(x,y)$ is _____ or _____, respectively, if it can be expressed in the form $f(x,y) = g(x)h(y)$ or $f(x,y) = g(x) + h(y)$. (2 words each)

637) If $a \le f(x) \le b$ for all x (in its #630), then $f(x)$ is said to be _____.

638) If a is $-\infty$ or b is ∞ in #637, then $f(x)$ is said to be _____.

639) A _____ #629 takes distinct and separate values.

640) A _____ #629 can take any real value within some interval.

641) $f(x)$ has a _____ at $x = a$ if $f(a)$ is not defined or if $f(x)$ doesn't approach the same value at $x = a$ from both the left and right sides.

642) $f(x)$ has a _____ at $x = a$ if there is not a unique tangent at $f(a)$.

643) A _____ or _____ of $f(x)$ is a value of x for which $f(x) = 0$.

644) $f(x)$ defined below on the left is an example of a _____. (2 words)

645) $g(x)$ defined below on the right is called the _____. (2 words)

$$f(x) = \begin{cases} 0, & \text{if } x < 0 \\ 1, & \text{if } x > 0 \end{cases} \quad , \quad g(x) = \begin{cases} -1, & \text{if } x < 0 \\ 0, & \text{if } x = 0 \\ 1, & \text{if } x > 0 \end{cases}$$

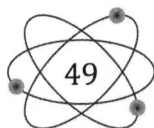

646) In $f(x)$, $g(y)$, and $h(p, q)$, the variables x, y, p, and q are called _____.

647) If $f(x)$ is either non-decreasing or non-increasing, it is said to be _____.

648) If $f(x)$ is _____ #647, it is either increasing or decreasing. (In this case, the prefix "non" from #647, which allows for a period of being constant, does not apply.)

649) If $f(x)$ has the form $axy + bx + cy + d$, it is said to be _____.

650) e^x is called the _____. (2 words)

651) The inverse of #650 is called a _____. It basically asks, "Which power does a given base need to be raised to in order to equal a given argument?"

652) The _____ is the time it takes for ae^{-kt} to be reduced to half of its initial value. (hyphenated)

653) $\frac{e^x + e^{-x}}{2}$ and $\frac{e^x - e^{-x}}{2}$ are examples of _____. (2 words)

654) _____ refers to a #629 that serves as its own inverse.

655) The _____ has a real axis and an imaginary axis. (2 words)

656) $x - iy$ is the _____ of $x + iy$. (2 words)

657) $\sqrt{x^2 + y^2}$ is the _____ of $x + iy$.

658) When we write $x + iy = re^{i\theta}$, we call θ the _____. (2 words)

659) A _____ has the form $a + ib + jc + kd$.

660) $a + ib$ is called a _____ integer if a and b are each integers.

661) A _____ is a value that $f(x)$ approaches as x tends to a particular value.

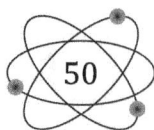

662) $f(x)$ is said to _____ if it approaches zero as x tends to a particular value.

663) An _____ value of a time-dependent variable is the value that it has at an exact moment in time.

664) A small increase in the value of a variable is called an _____.

665) The interval $a - \delta < x < a + \delta$ is referred to as the _____ of $x = a$.

666) A person might be _____ if that person is afraid of certain numbers like 13 and the number of this exercise.

667) If the right and left #661's of $f(x)$ differ, the difference is called a _____.

668) A problem is termed _____ if a small change in x causes a large change in $f(x)$. (hyphenated)

669) A _____ divides an interval into multiple subintervals.

670) $\frac{dy}{dx}$ is a _____ of y with respect to x.

671) $\frac{d^2y}{dx^2}$ is a _____ #670.

672) $\frac{d^3y}{dx^3}$ and $\frac{d^4y}{dx^4}$ are _____ #670's. (hyphenated)

673) _____ means to find the #670 of $f(x)$.

674) $f(x)$ is said to be _____ at $x = a$ if the #670 of $f(x)$ exists at $x = a$.

675) A complex #629 is said to be _____ or _____ if it is #674 at every point in its #630.

676) A curve is _____ if it is everywhere #674 and its #670 is continuous.

677) $\int f(x)\,dx$ is an _____ and $\int_{x=a}^{b} f(x)\,dx$ is a _____.
(2 words each)

678) $\int f(x)\,dx$ may also be called an _____. (sometimes hyphenated)

679) The process of finding #677 is termed _____.

680) In #677, $f(x)$ is called the _____.

681) The answer to #678 includes an _____ of #679. (2 words)

682) The variable x in $\int_{x=a}^{b} f(x)\,dx$ is called a _____ variable because the answer would be the same if it were changed to another symbol, such as y or t.

683) The _____ sum approximates #682 with a sum over small intervals. As the intervals become infinitesimal, the value of the sum approaches the value of the integral.

684) The _____ rule allows you to determine $\frac{dy}{dx}$ when you know $\frac{dy}{du}$ and $\frac{du}{dx}$.

685) An example of a _____ problem is when you are given $\frac{dx}{dt}$, but need to find $\frac{dy}{dt}$ using the relationship between x and y. (2 words)

686) _____ points (also called _____ points) are extreme values such as minima and maxima.

687) The points described in #686 can be found by applying the _____ test.

688) A _____ (or local) minimum/maximum is lower/higher than the points to its sides. It is not necessarily the lowest/highest point on the entire graph.

689) An _____ (or global) minimum/maximum is the lowest/highest point on a specified interval.

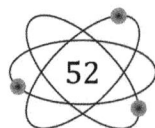

690) A point of _____ is where the concavity changes.

691) A _____ point is where a graph changes direction. It is yet another word for #688.

692) An _____ problem often involves maximizing or minimizing a function.

693) A _____ restricts the possible answers to #692.

694) _____ expresses an integral in terms of vdu rather than udv. (3 words)

695) Rewriting $\frac{1}{x^2-2x-3}$ as $\frac{A}{x-3} + \frac{B}{x+1}$ is called _____. (2 words)

696) An example of _____ is beginning with $x^2y^2 + xy + 1 = 0$ to obtain $x^2 2y\frac{dy}{dx} + 2xy^2 + x\frac{dy}{dx} + y = 0$. (2 words)

697) $\int_{x=0}^{\infty} e^{-x}\, dx$ and $\int_{x=0}^{1} \frac{dx}{\sqrt{x}}$ are examples of _____. (2 words)

698) A _____ of $f(x,y) = 3x^2y^2$ with respect to x is $6xy^2$. (2 words)

699) The _____ operator in Cartesian coordinates consists of three terms, each of which combines a unit vector with the corresponding #698.

700) The _____ operator has #699 act on a scalar field.

701) The _____ operator finds the dot product of #699 with a vector field.

702) The _____ operator finds the cross product of #699 with a vector field.

703) The _____ operator has #701 act on #700 for a scalar function.

704) The _____ represents the instantaneous rate of change in the direction of a curve (at a particular point along the curve).

705) _____ is a word that describes math with multiple independent variables.

706) $x_1, x_2, x_3, \ldots, x_n$ is a _____. $x_1 + x_2 + x_3 + \cdots + x_n$ is a _____.

707) A _____ is a <the first answer to #706> where x_n is related to x_{n-1} by a rule.

708) In the expression shown to the right, n is called the _____. $\sum_{n=1}^{\infty} \dfrac{1}{n^2}$

709) The sign of an _____ #706 changes with each term.

710) The terms of an _____ #706 have a constant difference.

711) The terms of a _____ #706 have a constant ratio.

712) If you take the reciprocal of each term of #710, you get a _____ #706.

713) A _____ #706 has terms of the form $a_n x^n$.

714) An _____ #706 has terms of the form $\dfrac{a_n}{x^n}$ and approximates a function of x in the limit that x approaches infinity.

715) An _____ #706 has terms of the form $\dfrac{z^n}{n!}$.

716) A _____ #706 has terms of the form $\dfrac{f^n(a)}{n!}(x-a)^n$.

717) If you set $a = 0$ in #716, you get a _____ #706.

718) A _____ #706 has terms of the form $a_n \cos(nx)$ and $b_n \sin(nx)$.

719) $\dfrac{1}{1} + \dfrac{1}{2} + \dfrac{1}{3} + \dfrac{1}{4} + \dfrac{1}{5} + \cdots$ _____, but $\dfrac{1}{1} + \dfrac{1}{4} + \dfrac{1}{9} + \dfrac{1}{16} + \dfrac{1}{25}$ _____.

720) The _____ test uses $\dfrac{a_{n+1}}{a_n}$ to help determine cases similar to #719.

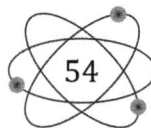

721) The _____ test compares a #706 to another #706 where the #720 is known.

722) The _____ is the #717 for $(1 + x)^p$, where $-1 < x < 1$. (2 words)

723) A _____ is an object with multiple indices, for which the scalar, vector, and matrix are (in a crude sense) special cases.

724) $x_n x_n$ has an implied _____, whereas $x_m x_n$ does not.

725) The _____ is unity when its two indices are equal and is zero when they differ. (2 words)

726) The _____ symbol has three indices that follow a cyclic order and is useful for forming the vector product in #723 form. (hyphenated)

727) The expression shown to the right involves _____ notation. $\prod\limits_{k=1}^{n} x_k$

728) An _____ is a quantity that remains constant under a transformation.

729) r and θ are 2D _____ coordinates.

730) r, θ, and φ are _____ coordinates.

731) ρ, φ, and z are _____ coordinates (notation varies to some extent).

732) The word _____ means outward. It describes r in #729 and #730.

733) The _____ angle is measured counterclockwise from the $+x$-axis after first projecting a vector down onto the xy plane. It is one of the angles in #730.

734) The squares of the _____ add up to unity, where the angles involved are measured from the x-, y-, and z-axes. (2 words)

735) xy, yz, and zx are referred to as the _____. (2 words)

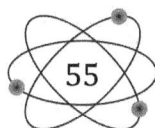

736) As a wheel rolls along the horizontal, a point on its edge traces out a _____.

737) A point on the edge of a wheel traces out a _____ if the wheel rolls along the inside of the circumference of a fixed circle.

738) The curve in #737 is called an _____ if the wheel's diameter is one-fourth of the fixed circle's diameter.

739) A point on the edge of a wheel traces out a _____ if the wheel rolls along the outside of the circumference of a fixed circle with the same diameter.

740) A heavy chain has the shape of a _____ if its ends are held fixed.

741) The _____ is the path of least time for a particle descending a height in a uniform gravitational field (neglecting resistive forces).

742) The _____ is the path for which the time it takes for a particle to descend to its lowest point is independent of where the particle starts in a uniform gravitational field.

743) A _____ is a polynomial equation of the second degree in Cartesian coordinates. (2 words)

744) The _____ of a parabola is a line segment passing through its focus which is perpendicular to the parabola's axis and connects points on the curve. (2 words)

745) Two _____ points on a sphere are connected by a diameter.

746) A half-revolution of an ellipse about a major or minor axis creates a _____.

747) #746 is called _____ if it is about a major axis and _____ if it is about a minor axis.

748) #746 is a special case of an _____ where two axes have the same length.

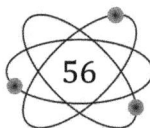

749) A half-revolution of a parabola about its axis creates a _____.

750) A _____ #749 is a doubly ruled surface that features a _____ (2 words).

751) An _____ #749 features a "nose cone."

752) A _____ is a #743 that may have one or two _____.

753) _____ is formed by revolving $y = \frac{1}{x}$ about the x-axis. (2 words)

754) _____ angle is a 3D generalization of the 2D concept of the angle.

755) The _____ is the SI unit of #754.

756) The rows (or columns) of a matrix are _____ if none of its rows (or columns) can be written as a linear combination of the others. (2 words)

757) The _____ of a matrix equals the number of its vectors that are #756.

758) The _____ matrix has 1's on its main diagonal and 0's elsewhere.

759) A _____ matrix has the same number of rows as it has columns.

760) A _____ matrix is a #759 where its off-diagonal elements equal zero.

761) The _____ of a matrix is found by replacing each A_{mn} with A_{nm}.

762) A _____ matrix equals its own #761.

763) $A_{nm} = -A_{mn}$ for an _____ matrix.

764) The product of a matrix and its _____ equals the #758 matrix.

765) The _____ for an element in a matrix is found by removing its column and row from the matrix and finding the determinant of what remains.

766) The _____ of a matrix is found by adding the elements of its main diagonal.

767) A _____ or _____ (hyphenated) matrix is a square matrix that is its own conjugate #761.

768) The #761 equals the #764 for an _____ matrix.

769) The _____ matrix for a system of linear equations includes the coefficients of the variables as well as an extra column for the constants on the right-hand side.

770) The _____ of a matrix is where the leading element in each row equals one, the leading one comes in a column to the right of the leading one from the previous row, and any rows with all zeros come below rows with nonzero elements. (2 words)

771) _____ is commonly used to transform a matrix to #770. (2 words)

772) The _____ for a square matrix is found by multiplying a constant by the identity matrix, subtracting this from the given matrix, and taking the determinant. (2 words)

773) In $A|a\rangle = \lambda|a\rangle$, λ is called the _____ and $|a\rangle$ is called the _____.

774) A square matrix is _____ if its determinant equals zero.

775) Two matrices are said to be _____ with regard to multiplication if the first matrix has the same number of columns as the second matrix has rows.

776) The _____ of a square matrix is found by first replacing each element with its cofactor and then taking the transpose.

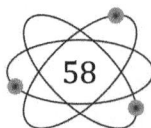

777) A matrix is _____ if it has square matrices along its main diagonal. (2 words)

778) A _____ matrix has zeros everywhere except for the main diagonal and its adjacent diagonals.

779) The _____ product is a row vector times a column vector. The _____ product is the column vector times a row vector (resulting in a $n \times n$ matrix).

780) The elements of the _____ matrix equal the first-order partial derivatives of a vector function.

781) An _____ equation involves x, y, $\frac{dy}{dx}$, $\frac{d^2y}{dx^2}$, etc. (2 words)

782) The _____ of #781 refers to its highest derivative.

783) A _____ equation involves x, y, t, $\frac{\partial y}{\partial x}$, $\frac{\partial y}{\partial t}$, $\frac{\partial^2 y}{\partial x^2}$, $\frac{\partial^2 y}{\partial t^2}$, $\frac{\partial^2 y}{\partial x \partial t}$, etc. (2 words)

784) The _____ are constraints added to #781 or #783 which specify values that variables must satisfy. (2 words)

785) A _____ problem is a #781 or #783 that includes #784. (2 words)

786) _____ #784's specify the values of the solution, while _____ #784's specify the values of the derivative of the solution.

787) $\frac{dy}{dx} + yP(x) = Q(x)$ has the form of a _____ #781. (3 words, hyphen)

788) For #787, $e^{\int P(x)dx}$ is called an _____. (2 words)

789) $a\frac{d^2y}{dx^2} + b\frac{dy}{dx} + c = f(x)$ is a _____ #781 with _____ (2 words)

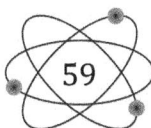

59

790) $\frac{dy}{dx} = f(x)g(y)$ has the form of a _____ #781. (3 words, hyphen)

791) If $f(kx, ky) = f(x, y)$ for all k, then $\frac{dy}{dx} = f(x, y)$ is a _____ #781. (3 words, hyphen)

792) $\frac{dy}{dx} = -2y$ and $\frac{dy}{dx} + y = 1$ are _____ #781's, but $\frac{dy}{dx} = -2x$ is not.

793) The _____ or _____ solution to a #781 with #782 equal to n has n arbitrary constants of integration.

794) A _____ solution to a #781 assigns values to the arbitrary constants.

795) The _____ equation has the form $ar^n + br^{n-1} + cr^{n-2} + dr^{n-3} + \cdots = 0$ for a #789 with #782 equal to n.

796) $dz = \frac{\partial z}{\partial x} dx + \frac{\partial z}{\partial y} dy$ is called an _____ if $z = z(x, y)$. (2 words)

797) $\frac{dA}{dt} = A - B$ and $\frac{dB}{dt} = A + B$ is an example of _____ #781's.

798) $r_{n+2} = r_{n+1} + r_n$ is a _____ relation (or _____ equation).

799) In _____ oscillations, the amplitude decreases as time progresses.

800) #799 is called _____ if it just barely isn't able to oscillate. (2 words)

801) _____ refers to the nature of equilibrium.

802) _____'s method is a first-order approximation to a #781 which takes small steps along the tangent lines from the initial position.

803) A _____ method uses midpoints to improve upon #802. (hyphenated)

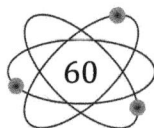

804) _____'s method uses $x_{n+1} = x_n - \frac{f(x_n)}{f'(x_n)}$ to find the roots of a function.

805) The _____ method for finding roots divides the interval in half and requires $f(a)$ and $f(b)$ to have opposite signs.

806) _____ refers to the process of finding area or a definite integral numerically.

807) _____'s rule uses parabolic arcs instead of line segments to approximate a definite integral numerically.

808) The _____ rule uses the straight line segments mentioned in #807.

809) _____ are the solutions to $\frac{d}{dx}(1-x)^2\frac{dy}{dx} = -cy$. (2 words)

810) _____ are the solutions to $x^2\frac{d^2y}{dx^2} = (c^2 - x^2)y$. (2 words)

811) The _____, defined by an integral, generalizes the factorial function to complex numbers (but excludes negative integers). (2 words)

812) The _____ generalizes $\binom{n}{k}$ similar to how #811 generalizes the factorial function. (2 words)

813) The _____ is defined by the sum shown to the right, provided that the real part of z is greater than unity. (2 words)

$$\sum_{n=1}^{\infty} \frac{1}{n^z}$$

814) The integral shown to the right is a kind of _____ integral.

$$\int_{\theta=0}^{t} \frac{d\theta}{\sqrt{1 - k^2 \sin^2\theta}}$$

815) The _____ is an idealized function that equals zero whenever its argument is nonzero and has an integral over all real numbers equal to unity. (3 words)

816) The integral shown to the right defines a _____. (2 words)

$$F(t) = \int_{x=-\infty}^{\infty} f(x)e^{ixt}\,dx$$

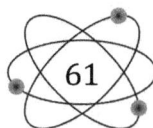

817) The integral shown to the right defines a _____. (2 words)

$$F(z) = \int_{t=0}^{\infty} f(t)e^{-xt}\, dt$$

818) Field lines diverge from _____ and converge toward _____.

819) A _____ refers to a small change in one or more parameters.

820) A map between two objects in category theory is called a _____.

821) A _____ method makes an approximation based on repeated sampling. (2 words)

822) A topological space that is locally Euclidean is called a _____.

823) The _____ (in a rough sense) gives the distance between two points in a given space. (2 words)

824) The _____ relates spacetime coordinates for relative observers for the case of constant velocity. (2 words)

825) The (nonrelativistic) _____ equals kinetic energy minus potential energy.

826) The integral of #825 over time is called the _____.

827) _____ are often used with #825 when there are constraints. (2 words)

828) The _____ formulation transforms the #825 formulation into a set of twice as many equations, but which are first-order instead of second-order.

829) A _____ is the integral of a function over a closed path in the complex plane. (2 words)

830) When performing a #829 around a pole, the nonzero part is called a _____.

831) A _____ space is a vector space where the norm is $\|x\| = \sqrt{\langle x, x \rangle}$.

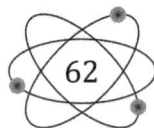

2 Number Challenge

1) How many dimensions of space are clearly evident in our universe?

2) Which number is neither positive nor negative?

3) Which number is considered unity?

4) What is π rounded to three decimal places?

5) Determine 10^6 without using a calculator.

6) Determine 2^8 without using a calculator.

7) Determine $\sqrt{9}$ without using a calculator.

8) Determine $\sqrt[3]{64}$ without using a calculator.

9) Estimate $\sqrt{2}$ to three decimal places.

10) Estimate $\sqrt{3}$ to three decimal places.

11) What is Euler's number, e, rounded to three decimal places?

12) What is Euler's constant, γ, rounded to three decimal places?

13) What is the golden ratio rounded to three decimal places?

14) Add the numbers 1 thru 60 without using a calculator or brute force.

15) Write $\frac{5}{8}$ as a percent.

16) Write $\frac{4}{11}$ as a repeating decimal.

17) Convert 30 from a decimal number to a binary number.

18) Convert 101,101 from a binary number to a decimal number.

19) Convert 100 from a decimal number to a hexadecimal number.

20) Convert 7D from a hexadecimal number to a decimal number.

21) How many millimeters equate to one centimeter?

22) How many centimeter equate to one kilometer?

23) How many cubic centimeters equate to one liter?

24) How many seconds equate to one day?

25) How many feet equate to one mile?

26) How many inches equate to one yard?

27) How many square feet equate to one square yard?

28) How many centimeters equate to one inch?

29) How many feet equate to one furlong?

30) How many miles equate to one nautical mile?

31) How many pounds equate to one ton in the United States?

32) How many pounds equate to one (imperial) ton in the United Kingdom?

33) How many pounds equate to one tonne (or metric ton) in the United Kingdom?

34) How many square yards equate to one acre?

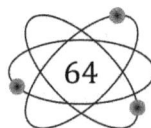

35) Convert 94 from an ordinary number to a Roman numeral.

36) Convert CMXLV from a Roman numeral to an ordinary number.

37) How many faces, edges, and vertices does a cube have?

38) How many faces, edges, and vertices does a tetrahedron have?

39) How many faces, edges, and vertices does a dodecahedron have?

40) How many degrees correspond to a right angle?

41) How many degrees correspond to a full circle?

42) In *The Hitchhiker's Guide to the Galaxy*, what is the answer to the universe?

43) What is the ratio of the sides of a 45° right triangle?

44) What is the ratio (from shortest to longest) of the sides of a 30°-60°-90° triangle?

45) What is the interior angle of an equilateral triangle in degrees?

46) What is the interior angle of a regular pentagon in degrees?

47) What is the bond angle of a tetrahedral molecule in degrees to one decimal place?

48) What is the dihedral angle of a regular tetrahedron in degrees to one decimal place?

49) How many balls can be arranged to touch a central ball of the same size in 3D space?

50) How many radians correspond to a full circle?

51) How many steradians correspond to a full sphere?

52) How many degrees correspond to one radian to one decimal place?

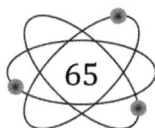

53) If x is nonzero, what is x^0?

54) What are 0!, 1!, and 5!?

55) When you expand $(x + y)^7$, what are the coefficients?

56) Determine $\sin 0°$, $\sin 30°$, $\sin 45°$, $\sin 60°$, and $\sin 90°$ without using a calculator.

57) Determine $\cos 0°$, $\cos 30°$, $\cos 45°$, $\cos 60°$, and $\cos 90°$ without using a calculator.

58) Determine $\tan 0°$, $\tan 30°$, $\tan 45°$, $\tan 60°$, and $\tan 90°$ without using a calculator.

59) Determine $\cos 120°$, $\sin 135°$, and $\tan 300°$ without using a calculator.

60) Determine $\sec 60°$, $\csc 240°$, and $\cot 150°$ without using a calculator.

61) Determine $\sin 15°$ without using a calculator.

62) Determine $\sin^{-1}\left(\frac{1}{2}\right)$, $\cos^{-1}\left(-\frac{\sqrt{2}}{2}\right)$, and $\tan^{-1}(-1)$ without using a calculator.

63) What are $\log_{10} 100$ and $\ln 1$ exactly, and what is $\ln 2$ to three decimal places?

64) Evaluate $e^{i\pi} + 1$.

65) What is $\sqrt{i}$?

66) How many dimensions are there in string theory, superstring theory, and M theory?

67) How many cubes, faces, edges, and vertices does a tesseract (a 4D hypercube) have?

68) For which value of x is $\frac{1}{x^3} - \frac{1}{x^2}$ a minumum?

69) Evaluate $\displaystyle\int_{x=-\infty}^{\infty} e^{-x^2}\, dx$.

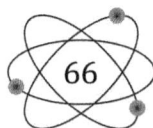

66

3 Symbol Challenge

1) What is the standard symbol for addition?

2) What are two common ways to indicate subtraction? (One is used in accounting.)

3) What are three common ways to indicate multiplication with numbers?

4) How is multiplication commonly shown with variables?

5) What are three common ways to indicate division?

6) What are two common ways to express a fraction?

7) Which standard symbol is used to indicate a ratio between two numbers?

8) What are two common ways to express A is to B as C is to D?

9) Which standard symbol is used to indicate a percent?

10) What are three symbols used as a decimal separator in different parts of the world?

11) What are three symbols used as a thousands separator in different parts of the world?

12) What is the standard symbol for equality?

13) What are five common symbols for inequalities?

14) Which symbol is used when two quantities are approximately equal?

15) Which symbol means "on the order of"?

16) Which symbol means "proportional to"?

17) What are two common ways to indicate a square root?

18) What are three common ways to bracket numbers or expressions?

19) How are absolute values indicated?

20) Which symbol means "and so on"?

21) Which symbols represent 1, 5, 10, 50, 100, 500, and 1000 in Roman numerals?

22) Which symbol represents the set of real numbers?

23) Which symbol represents the set of complex numbers?

24) Which symbol represents United States dollars?

25) What are two ways to indicate cents in the United States?

26) Which symbol represents pounds sterling in the United Kingdom?

27) Which symbol represents euro in continental Europe?

28) Which symbol represents Japanese yen?

29) Which symbol represents the Indian rupee?

30) Which symbol is sometimes referred to as the octothorpe, hash sign, or pound sign?

31) Which symbol is used to indicate both positive and negative alternatives?

32) Which symbol represents infinity?

33) Which symbols represent 10, 11, 12, 13, 14, and 15 in hexadecimal notation?

34) Which lowercase Greek letter represents the number pi?

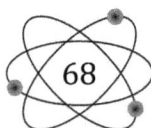

35) Which lowercase Greek letter is most commonly used to represent an angle?

36) What are two common variations of the lowercase Greek letter phi?

37) Which uppercase Greek letter represents a change in a quantity?

38) Which lowercase Greek letter represents a small change in a quantity?

39) Which uppercase Greek letter represents a sum?

40) Which uppercase Greek letter represents a product?

41) What are two common variations of the lowercase Greek letter epsilon?

42) Which lowercase Greek letter is commonly used to represent angular speed?

43) Which lowercase Greek letter is commonly used for the coefficient of friction?

44) Which lowercase Greek letter is referred to in a chi-squared distribution?

45) Which standard symbol is used to represent the Euler number?

46) Which symbol is commonly used to represent Euler's constant?

47) Which standard symbol is used for degrees?

48) Which symbols are used to divide degrees into arc minutes and arc seconds?

49) How is the distance between points A and B represented?

50) How is the finite line segment joining points A and B represented?

51) How is the ray extending from point A and passing through point B represented?

52) How is the infinite line passing through points A and B represented?

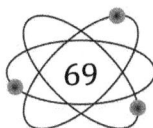

53) Which symbol is commonly used to indicate that letters or numbers represent an angle?

54) Which symbol is commonly used to indicate a right angle?

55) Which symbol is commonly used to indicate the angular measure of an angle?

56) Which symbol is commonly used to indicate that letters form a triangle?

57) Which symbol is commonly used to indicate that letters form a right triangle?

58) Which symbol means "parallel"?

59) Which symbol means "perpendicular to"?

60) Which symbol means "not parallel"?

61) Which symbol represents the congruence of two geometric figures?

62) Which symbol represents the similarity of two geometric figures?

63) Which standard symbols represent the slope and y-intercept of a straight line?

64) Which symbols are commonly used for length, width, height, and depth?

65) Which symbols are commonly used for the base and altitude of a triangle?

66) Which symbols are commonly used for area, perimeter, and volume?

67) Which symbols are commonly used for radius, diameter, and circumference?

68) Which symbol is commonly used for radius of gyration?

69) Which symbol is commonly used for arc length?

70) Which symbol is commonly used for focal length?

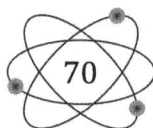

71) Which symbols are commonly used for the semimajor and semiminor axes?

72) Which symbol is commonly used for eccentricity?

73) Which symbol (**not** an abbreviation) commonly appears at the end of a proof?

74) Which symbol means "is defined as"?

75) Which symbol means "implies that"?

76) Which symbol means "if and only if"?

77) Which symbol means "therefore"?

78) Which symbol means "because"?

79) Which symbol means "such that"?

80) Which symbol means "there exists"?

81) Which symbol means "for any" or "for all"?

82) Which symbol means "belongs to" or "is an element of"?

83) What are the symbols for union and intersection?

84) What are the symbols for conjunction and disjunction?

85) What is the symbol for logical negation, known as the not sign (**not** the minus sign)?

86) What are the symbols for subset or superset and for proper subset or proper superset?

87) Which symbol represents the empty set?

88) Which symbol is used to indicate the average value of a variable?

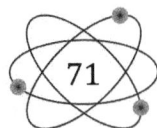

89) Which symbol represents a factorial?

90) What is a common way to write the binomial coefficient, read as "n choose m"?

91) Which symbols represent the possible outcomes of a coin flip?

92) Which symbols represent the possible entries of a truth table?

93) Which symbols enclose an array to represent a matrix?

94) Which symbols enclose an array to represent a determinant?

95) Which symbols represent the adjacent, opposite, and hypotenuse on a unit circle?

96) What are two common ways to indicate that a quantity is a vector?

97) What is a common way to indicate that a quantity is a unit vector?

98) Which symbols are commonly used for Cartesian unit vectors?

99) What are two common ways to indicate the magnitude of a vector?

100) What are two common methods of distinguishing between initial and final values?

101) Which symbols represent distance, rate, and time?

102) Which symbol is commonly used to represent speed?

103) Which symbols commonly represent the position and displacement vectors?

104) Which symbols commonly represent velocity and acceleration?

105) Which symbols commonly represent angular velocity and angular acceleration?

106) Which symbols commonly represent frequency and period?

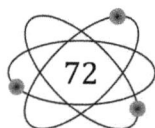

107) Which symbol commonly represents wavelength?

108) Which symbols (that aren't letters) are sometimes used to indicate feet or inches?

109) What are the symbols for the SI units of length, mass, and time?

110) What are the symbols for the dimensions of length, mass, and time?

111) What are the symbols for the SI units of velocity and acceleration?

112) What are the symbols for the SI units of frequency and period?

113) What are the symbols for the SI units of moment of inertia?

114) What are the symbols for the SI units of force, work, and power?

115) What are the symbols for the SI units of momentum and angular momentum?

116) What are the symbols for the SI units of electric field and magnetic field?

117) What are the symbols for the SI units of heat and temperature?

118) What are the symbols for the SI units of density?

119) What are the symbols for the SI units of entropy?

120) What are the symbols for the SI units of current, resistance, and potential difference?

121) How are the scalar product and vector product represented?

122) How is the natural logarithm of a number indicated?

123) How is the real or imaginary part of a complex number indicated?

124) Which symbol represents the complex conjugate of a quantity?

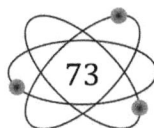

125) Which symbols represent Cartesian coordinates?

126) Which symbols represent 2D polar coordinates?

127) Which symbols represent spherical coordinates?

128) Which symbols represent cylindrical coordinates?

129) How is a limit represented?

130) Which symbol means "approaches"?

131) How is a limit represented if it is approached from the left or right?

132) Which symbols mean "much less than" or "much more than"?

133) What are two common ways to indicate a derivative?

134) What is the standard symbol for integration?

135) How is a partial derivative represented?

136) Which symbols represent the gradient, divergence, and curl operators?

137) Which symbols represent the Laplacian and d'Alembertian operators?

138) Which symbols represents arrows pointing into or out of the page?

139) Which symbol represents an inverse?

140) Which symbol represents the transpose of a matrix?

141) What is the bra-ket notation for the inner product between vectors?

142) Which symbol represents the Hermitian conjugate in bra-ket notation?

4 Formula Challenge

1) Which formula illustrates the additive identity?

2) Which formula illustrates the multiplicative identity?

3) Which formula illustrates the commutative property of addition?

4) Which formula illustrates the commutative property of multiplication?

5) Which formula illustrates the associative property of addition?

6) Which formula illustrates the associative property of multiplication?

7) Which formula illustrates the distributive property?

8) Which formula illustrates the property of the additive inverse?

9) Which formula illustrates the property of the multiplicative inverse?

10) What is the formula to compute simple interest?

11) What is the formula for compound interest?

12) What is the formula associated with the f.o.i.l. method?

13) What is the formula for the difference of squares in algebra?

14) What is the quadratic formula to solve an equation of the form $ax^2 + bx + c = 0$?

15) What is the formula for multiplying different powers of the same base?

16) What is the formula for dividing different powers of the same base?

17) Which formulas are associated with Cramer's rule?

18) What is Pell's equation?

19) What is the formula for the Pythagorean theorem?

20) What is the distance formula in 3D space?

21) What is the midpoint formula in 3D space?

22) What is the section formula in 3D space?

23) What is the slope formula for a line in the xy plane?

24) What is the slope-intercept form of the equation for a line in the xy plane?

25) What is the point-slope form of the equation for a line in the xy plane?

26) What is the two-intercept form of the equation for a line in the xy plane?

27) What is the formula that relates the slopes of perpendicular lines in the xy plane?

28) What is the constant rate equation?

29) What are the formulas for the perimeter and area of a rectangle?

30) Which formula relates radius to diameter?

31) What are the formulas for the circumference and area of a circle?

32) What is the formula for the arc length of a circular arc?

33) What is the formula for the area of a triangle?

34) What is the formula for the area of a parallelogram?

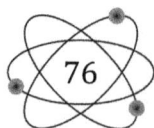

35) What is the formula for the area of a rhombus in terms of its diagonals?

36) What is the formula for the area of a trapezoid?

37) What is the formula for the area of a regular hexagon?

38) What is the formula for the area of a regular octagon?

39) What is the formula for the area of an ellipse?

40) What is the equation for a circle centered about the origin in the xy plane?

41) What is the equation for a parabola that is symmetric about the y-axis?

42) What is the equation for a hyperbola with its transverse axis along the x-axis?

43) What is the equation for an ellipse that is symmetric about the x- and y-axes?

44) What is the equation for a sine wave oscillating about the x-axis?

45) What is the equation for an Archimedean spiral?

46) What is the equation for an equiangular spiral?

47) What are the equations for parabolic and hyperbolic spirals?

48) What is the ratio formula for the golden section (the equation to solve, **not** the answer)?

49) What is the formula for the location of the centroid of a triangle in 3D space?

50) What is the formula associated with Ceva's theorem (for triangles)?

51) What is the formula associated with the intersecting chords theorem?

52) What is the formula associated with the intersecting secants theorem?

53) What are the formulas for the surface area and volume of a cube?

54) What are the formulas for the surface area and volume of a cuboid?

55) What are the formulas for the surface area and volume of a parallelepiped?

56) What are the formulas for the surface area and volume of a sphere?

57) What are the formulas for the surface area and volume of a right-circular cylinder?

58) What are the formulas for the surface area and volume of a cone?

59) What are the formulas for the surface area and volume of a regular tetrahedron?

60) What are the formulas for the surface area and volume of a single-holed ring torus?

61) What is the equation for a plane?

62) What are the equations for a line in 3D space?

63) What are the parametric equations for a helix winding around the z-axis?

64) What is the equation for a sphere centered about the origin?

65) What is the equation for an ellipsoid that is symmetric about the Cartesian axes?

66) What is Euler's formula relating the faces, vertices, and edges of a convex polyhedron?

67) What is the formula for the determinant of a 2×2 matrix?

68) What is the formula for the determinant of a 3×3 matrix?

69) What is the formula for the arithmetic mean?

70) What is the formula for the geometric mean?

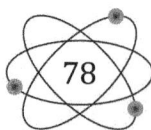

71) What is the formula for weighted average?

72) What is the formula for the range of data values?

73) What is the formula for the interquartile range?

74) What is the formula for the standard deviation?

75) What is the formula for the number of permutations?

76) What is the formula for the number of distinct permutations for repeated objects?

77) What is the formula for the number of permutation of n objects taken r at a time?

78) What is the combination formula?

79) What is the formula for Bayes' theorem (relating to probability)?

80) What are the formulas for percent error and percent difference?

81) What is the formula for chi-squared?

82) What is the equation for the Gaussian (normal) distribution?

83) What is the formula for the sine of an angle in a right triangle?

84) What is the formula for the cosine of an angle in a right triangle?

85) What is the formula for the tangent of an angle in a right triangle?

86) What is the formula for the cosecant of an angle in a right triangle?

87) What is the formula for the secant of an angle in a right triangle?

88) What is the formula for the cotangent of an angle in a right triangle?

89) Which formula relates the three basic trig functions (sine, cosine, and tangent)?

90) Which three common trig identities are forms of the Pythagorean theorem?

91) What is the sum or difference of angles formula for sine?

92) What is the sum or difference of angles formula for cosine?

93) What is the sum or difference of angles formula for tangent?

94) What is the double angle formula for sine?

95) What are three forms of the double angle formula for cosine?

96) What is the double angle formula for tangent?

97) What is the half-angle formula for sine?

98) What is the half-angle formula for cosine?

99) What is the formula for the law of cosines?

100) What is the formula for the law of sines?

101) What is the formula for the law of tangents?

102) What is Heron's formula, also known as Hero's theorem (regarding triangles)?

103) Which formula expresses a vector in terms of unit vectors in 3D space?

104) Which formula expresses the magnitude of a vector in terms of its components?

105) What are the formulas for the components of a vector in the xy plane?

106) What is the formula for the direction (angle) of a vector in the xy plane?

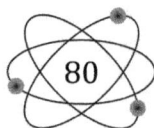

107) What is the formula for the position vector in 3D space?

108) Which calculus-based formula relates the velocity to the position vector?

109) Which calculus-based formulas relate the acceleration to the velocity and position?

110) What are the three formulas for one-dimensional uniform acceleration?

111) Which formula relates frequency to period?

112) Which formulas relate angular frequency to frequency and period?

113) Which formulas relate wavelength to wave speed?

114) What are two common formulas for the scalar product?

115) Which formula for the vector product involves a determinant?

116) What is the formula for the magnitude of the vector product?

117) What is the formula for the work done by a force?

118) What is the (vector) formula for torque?

119) Which (vector) formula relates velocity to angular velocity?

120) What is the formula for the vector triple product?

121) What is the formula for the scalar triple product?

122) What is the Jacobi identity (regarding vectors)?

123) Which formula relates the direction cosines in 3D space?

124) What are the formulas for a rotation of the coordinate axes in the xy plane?

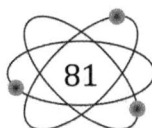

125) What is the formula to find the angle between two lines in the xy plane?

126) What is the formula to find the angle between two vectors in 3D space?

127) What is the formula for the logarithm of a product?

128) What is the formula for the logarithm of a reciprocal?

129) What is the formula for the logarithm of a quotient?

130) What is the formula for the logarithm of an argument raised to a power?

131) What is the formula for the natural logarithm of an exponential function?

132) What is the change of base formula for logarithms?

133) What is the formula for hyperbolic cosine?

134) What is the formula for hyperbolic sine?

135) What is the formula for hyperbolic tangent?

136) What is the formula for linear interpolation?

137) Which formula involving a limit serves to define what a derivative is?

138) Which formula equates Euler's constant to a limit?

139) What is the formula for the derivative of a power of a variable?

140) What are the formulas for the derivatives of sine, cosine, and tangent?

141) What are the formulas for the derivatives of cosecant, secant, and cotangent?

142) What are the formulas for the derivatives of arcsine, arccosine, and arctangent?

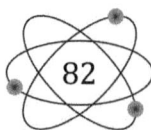

143) What is the formula for the derivative of the exponential function?

144) What is the formula for the derivative of a natural logarithm?

145) What is the formula for the derivative of a positive constant raised to a variable power?

146) What are the formulas for the derivatives of the three basic hyperbolic trig functions?

147) What is the formula for the product rule?

148) What is the formula for the quotient rule?

149) What is the formula for the chain rule?

150) What is Leibniz's theorem (regarding calculus)?

151) What is the formula to find the area under a curve?

152) What is the formula for the antiderivative of a power (other than -1) of a variable?

153) What is the formula for the antiderivative of the reciprocal of a variable?

154) What are the formulas for the antiderivatives of sine and cosine?

155) What are the formulas for the antiderivatives of tangent and cotangent?

156) What are the formulas for the antiderivatives of cosecant and secant?

157) What is the formula for the antiderivative of the exponential function?

158) What is the formula for the antiderivative of a natural logarithm?

159) What is the formula for the antiderivative of a constant raised to a variable power?

160) What are the formulas for the antiderivatives of hyperbolic sine and cosine?

161) What is the formula for integration by parts?

162) What is the formula for arc length involving an integral for a curve in the xy plane?

163) What is the formula for the center of mass of a system of discrete particles?

164) What is the formula for the center of mass of a continuous 3D object?

165) What is the formula for the moment of inertia of a system of discrete particles?

166) What is the formula for the moment of inertia of a continuous 3D rigid body?

167) What is the formula for the binomial (series) expansion?

168) What is the formula for the probability density function for a continuous variable?

169) What is the formula for the gradient operator in Cartesian coordinates?

170) What is the formula for the divergence operator in Cartesian coordinates?

171) What is the formula for the curl operator in Cartesian coordinates?

172) What is the formula for the Laplacian operator in Cartesian coordinates?

173) What is the formula for the d'Alembertian operator in Cartesian coordinates?

174) What is the formula for a complex number in terms of its real and imaginary parts?

175) What is the formula for the complex conjugate of a complex number?

176) What is the formula for the modulus-squared of a complex number?

177) What is the polar form of a complex number?

178) What is Euler's formula (regarding complex numbers)?

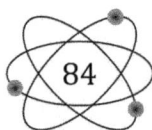

179) What is the formula associated with De Moivre's theorem?

180) What is the formula for the inverse of a complex number?

181) Which formulas are associated with a quaternion?

182) What is the formula for Cauchy's integral theorem?

183) What is the characteristic equation (regarding matrices)?

184) What is the eigenvalue equation (characteristic of the eigenvalue problem)?

185) What is the formula for the inverse of a 2×2 matrix?

186) What are the formulas for the commutator and anticommutator of matrices?

187) What is the Jacobi identity (regarding matrices)?

188) What is the formula for the curvature (regarding calculus)?

189) What are the Lotka-Volterra predator-prey equations?

190) What is the (integral) formula associated with Gauss's law?

191) What is the (integral) formula associated with Stokes' theorem?

192) What is the (integral) formula associated with Green's theorem?

193) What is Laplace's (differential) equation?

194) What is Stirling's formula?

195) Which formulas express Cartesian coordinates in terms of 2D polar coordinates?

196) Which formulas express 2D polar coordinates in terms of Cartesian coordinates?

197) Which formulas express Cartesian coordinates in terms of spherical polar coordinates?

198) Which formulas express spherical unit vectors in terms of Cartesian unit vectors?

199) What is the formula for the differential volume element in spherical coordinates?

200) What is the formula for the gradient operator in spherical coordinates?

201) What is the formula for the divergence operator in spherical coordinates?

202) What is the (determinant) formula for the curl operator in spherical coordinates?

203) What is the formula for the Lorentz (time dilation and length contraction) factor?

204) What are the formulas for the Lorentz transformation for relative velocity along x?

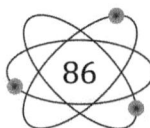

5 Pattern Challenge

1) 8, 15, 22, 29, ____, ____

2) 143, ____, 107, 89, ____, 53

3) C, E, G, ____, ____, M

4) x, w, ____, ____, t, s

5) b, ____, g, j, l, ____

6) ____, S, P, ____, J, G

7) 2, 3, 5, 8, 12, ____, ____

8) 256, 224, 194, 166, ____, ____

9) a, b, d, e, g, o, ____, ____

10) 3, 4, 7, 11, 18, ____, ____

11) 141, 87, 54, ____, 21, ____

12) 2, 3, 4, 9, 16, 29, 54, ____, ____

13) 3, 15, 75, 375, ____, ____

14) 2916, 972, ____, 108, ____, 12

15) 2, 4, 12, 48, 240, ____, ____

16) 46,080, 3840, 384, ____, 8, 2, ____

17) 1, 2, _____, _____, 120, 720, 5040

18) 9, 25, 49, 81, _____, _____

19) 4913, 2744, 1331, 512, _____, _____

20) 2, 9, 64, 625, _____, _____

21) I, ____, X, L, ____, D, M

22) ____, 11, 101, 111, 1001, 1011, 1101, _____

23) ee, ix, ne, ve, en, en, ne, ur, en, ____, ____

24) 11, 13, 17, 19, 23, ____, ____

25) 2, 9, 6, 13, 10, 17, ____, ____

26) 3, 9, 15, 45, 51, 153, _____, _____

27) 10, 25, 55, 115, 235, _____, _____

28) 6, 8, 10, 16, 14, 32, 18, 64, 22, _____, _____

29) 4, 3, 5, 6, 7, 9, 10, 12, 14, 15, 19, ____, ____

30) 62, 47, 55, 55, 48, 63, 41, 71, 34, 79, 27, ____, ____

31) 1, 2, 3, 6, 9, 54, 63, _____, _____

32) 1234, 1243, 1324, 1342, 1423, 1432, _____, _____

33) 2X, 3V, 4T, 5R, ____, ____

34) Math, kDqi, igNj, gjkK, Emhl, _____, _____

35) 8:00, 9:45, 11:30, 1:15, _____, _____

36) iv, ix, xvi, xxv, xxxvi, xlix, _____, _____

37) 2, 9, 28, 65, 126, 217, _____, _____

38) 1, 2, 4, 6, 10, 12, 16, 18, 22, ____, ____

39) 1, 4, 27, 256, 3125, _____, _____

40) III, VII, XIII, XIX, XXIX, XXXVII, XLIII, _____, _____

41) 7, E, 15, 1C, 23, 2A, 31, ____, ____

42) $\frac{1}{24}$, $\frac{1}{8}$, $\frac{3}{8}$, $1\frac{1}{8}$, $3\frac{3}{8}$, _____, _____

43) $\frac{1}{8}$, 0.25, 37.5%, _____, _____, 75%, $\frac{7}{8}$, 1, 112.5%

44) $\frac{1}{12}$, $\frac{1}{6}$, $\frac{1}{4}$, $\frac{5}{12}$, $\frac{2}{3}$, $\frac{13}{12}$, ____, ____

45) 0.5, 0.25, 0.125, 0.0625, 0.03125, _____, _____

46) 256, 384, 576, 864, _____, _____

47) $\frac{5}{2}$, $\frac{7}{3}$, $\frac{11}{5}$, $\frac{15}{7}$, $\frac{23}{11}$, $\frac{27}{13}$, ____, ____

48) $\frac{5}{6}$, $\frac{1}{3}$, 1, $\frac{1}{2}$, $\frac{3}{2}$, 1, 3, $\frac{5}{2}$, ____, ____

49) $\frac{5}{6}$, $\frac{3}{4}$, $\frac{2}{3}$, ____, $\frac{1}{2}$, ____, $\frac{1}{3}$

50) 3, 8, 24, 48, 120, 168, 288, 360, _____, _____

51) 362,880, 322,560, 282,240, 241,920, 201,600, 161,280, 120,960, _____, _____

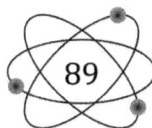

52)

1	3
3	0

2	4
8	4

5	7
35	28

8	10
80	70

18	

53)

6	3
15	5

12	12
36	3

45	15
60	4

100	25
175	7

180	
360	

54)

2	4
5	2

6	9
3	36

12	5
9	48

8	20
25	40

25	30
10	

55)

2	10
3	4

50	15
5	7

18	9
90	8

12	10
80	200

15	
20	100

56)

16	3
1	2

25	9
8	4

100	6
4	8

225	14
6	7

	20
10	15

57)

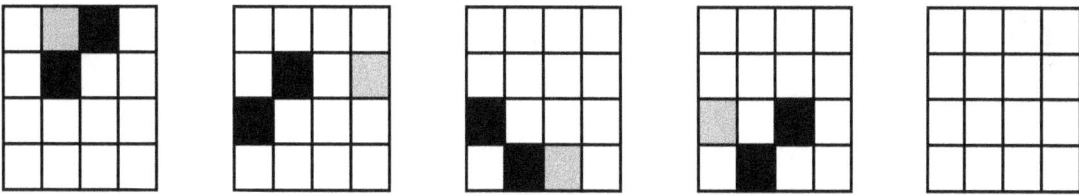

58)

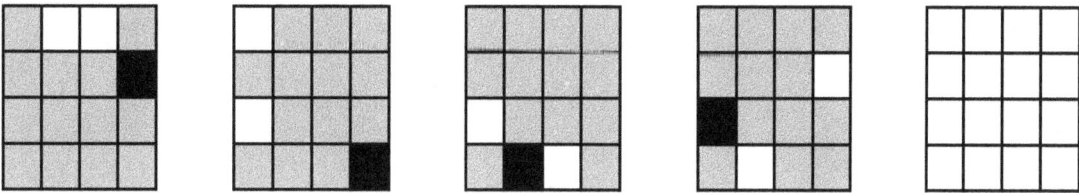

59)

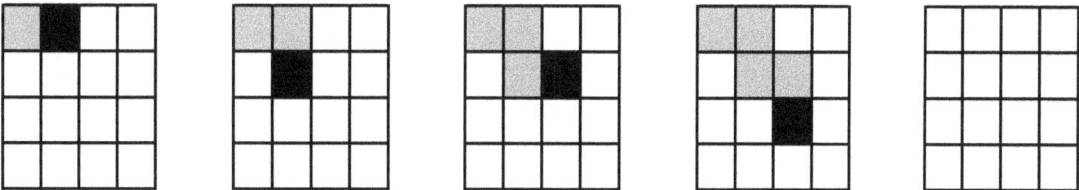

60)

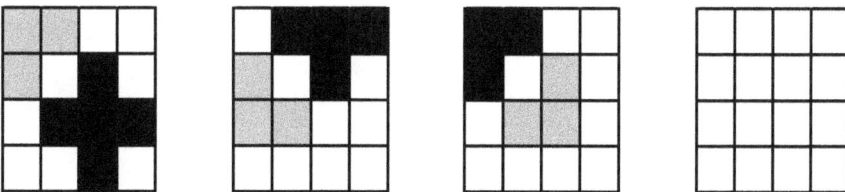

61)

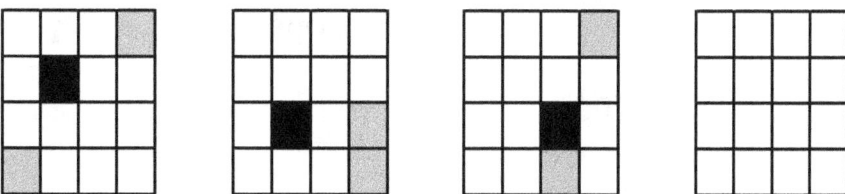

62)

63)

64)

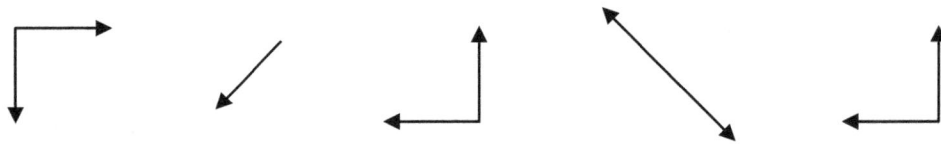

65)

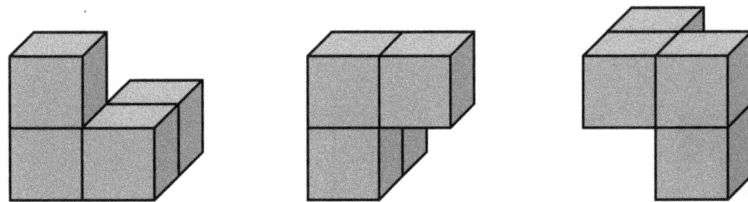

66)

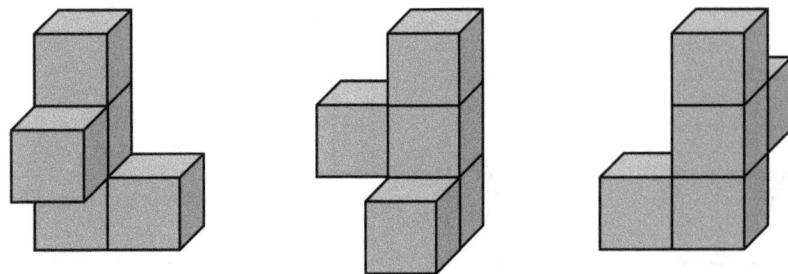

6 Test Your Math Skills

Note: These exercises are designed to be completed *without* the aid of a calculator.

1) Evaluate $-6 + 9 - 8 + 7 - (-5)$.

2) Evaluate $8 \times (-7)$, $-4 \times (-9)$, $-42 \div 6$, and $-72 \div (-9)$.

3) Evaluate 12^2, $(-3)^4$, $(-4)^3$, 5^1, 6^0, and 2^7.

4) Evaluate $16 - 4 \times 3 + 12 \div 2^2$.

5) Evaluate $18 \div (6 + 3) \times 4$.

6) Evaluate $\frac{24-12}{4+2}$.

7) Evaluate $\sqrt{49}$, $\sqrt{196}$, and $\sqrt{1,000,000}$ (in US notation).

8) Evaluate $\sqrt[3]{-64}$ and $\sqrt[4]{16}$ (for their real roots).

9) Determine the prime factorization of 90.

10) What is the greatest common factor of 72 and 108?

11) What is the least common multiple of 16 and 20?

12) Simplify $\sqrt{72}$ by factoring out all perfect squares.

13) Evaluate $2746 + 539$.

14) Evaluate $3241 - 437$.

15) Evaluate 869×7.

16) Evaluate 37×28.

17) Evaluate $1896 \div 24$.

18) Evaluate $1529 \div 42$ using a remainder.

19) Reduce $\frac{48}{60}$.

20) Convert $4\frac{3}{5}$ to an improper fraction.

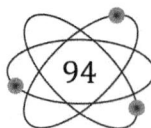

21) Evaluate $\frac{3}{4} + \frac{1}{6}$.

22) Evaluate $\frac{7}{8} - \frac{5}{12}$.

23) Evaluate $\frac{3}{8} \times \frac{2}{9}$.

24) Evaluate $\frac{4}{9} \div \frac{2}{3}$.

25) Express 4^{-1}, 3^{-2}, $\left(\frac{3}{4}\right)^2$, and $\left(\frac{2}{5}\right)^{-3}$ as fractions.

26) Evaluate $64^{2/3}$ (for real roots).

27) Evaluate $16^{-3/4}$ (for real roots).

28) Rationalize the denominator of $\frac{1}{\sqrt{3}}$.

29) Evaluate 10^5, 10^{-2}, and 10^{-4}.

30) Round 673.24519 to the nearest hundredth.

31) Evaluate $4.385 + 2.637$.

32) Evaluate $14.38 - 6.45$.

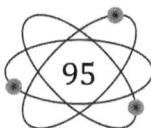

33) Evaluate 8.4×2.7.

34) Evaluate $5.6 \div 2.5$.

35) Convert 0.36 to a percent and to a reduced fraction.

36) Convert $\frac{5}{8}$ to a decimal and to a percent.

37) Convert 60% to a decimal and to a reduced fraction.

38) Convert $\frac{25}{33}$ to a repeating decimal.

39) Convert the repeating decimal $0.\overline{48}$ to a reduced fraction.

40) Order 157%, $1\frac{2}{3}$, 1.65, and $\frac{8}{5}$ from least to greatest.

41) How many girls attend a school with 135 students if the ratio of girls to boys is 5:4?

42) Convert 72 km/hr to m/s.

43) What is the average speed of a car that travels 510 km in 6 hours?

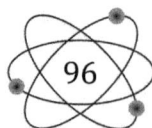

44) Traveling 35 km/hr, how far does a train travel in 4 hours?

45) How many minutes will it take to travel 24 km at a speed of 40 km/hr?

46) Solve $3x - 6 = 18$.   47) Solve $8 - 2x = 5x - 6$.

48) Simplify $\frac{x^9 x^3}{x^7 x^{-5}}$.

49) Expand $(2x^3)^4$.

50) Distribute $4x^3(3x^2 + 5x - 2)$.

51) Factor $8x^5 - 12x^3$.

52) Expand $(2x + 3)(4x - 5)$.

53) Expand $(3 - \sqrt{x})(3 + \sqrt{x})$.

54) Rationalize the denominator of $\frac{2}{\sqrt{2x}}$.

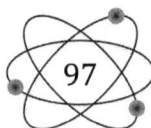

55) Solve $\dfrac{9}{x} = \dfrac{3}{2}$.

56) Solve $\dfrac{1}{3} - \dfrac{1}{x} = \dfrac{1}{12}$.

57) Describe the solution(s) to $x - 3 = x + 5$.

58) Describe the solution(s) to $3x + 4 = 2 + 3x + 2$.

59) Find all solutions to $x^2 - 4x = 0$.

60) Find all solutions to $2x^2 - 18 = 0$.

61) Find all solutions to $x^2 - 6x + 9 = 0$.

62) Find all solutions to $2x^2 + 9x = 18$.

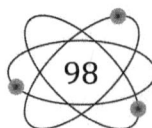

63) Solve the system $3x - 4y = 4$ and $5x + 2y = 50$.

64) Solve $2 - x < 7$.

65) Solve $9 + x > 1 + 3x$.

66) Determine θ. (Not drawn to scale.)

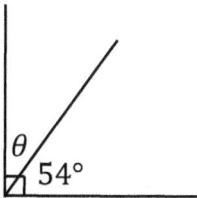

θ

$54°$

67) Determine θ. (Not drawn to scale.)

θ $28°$

68) Determine θ. (Not drawn to scale.)

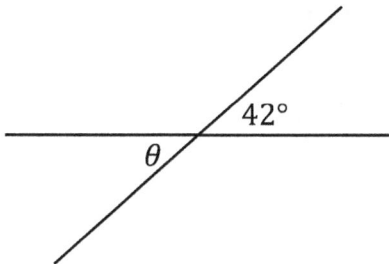

$42°$

θ

69) Determine θ. (Not drawn to scale.)

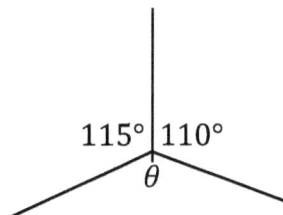

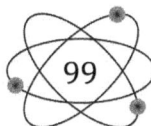

$115°$ $110°$

θ

99

70) Determine θ. (Not drawn to scale.)

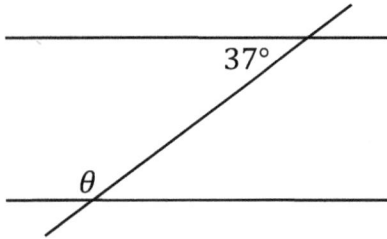

37°

θ

71) Determine θ. (Not drawn to scale.)

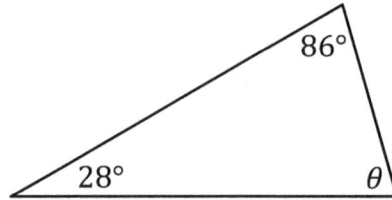

86°

28°

θ

72) Determine x for the right triangle.

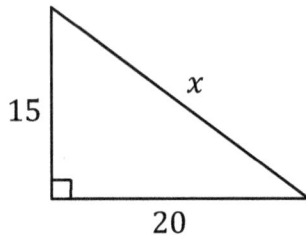

15

x

20

73) Determine x for the right triangle.

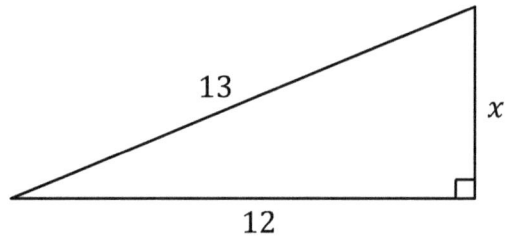

13

x

12

74) Determine x for the right triangle.

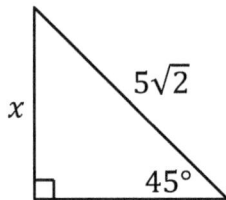

$5\sqrt{2}$

x

45°

75) Determine x for the right triangle.

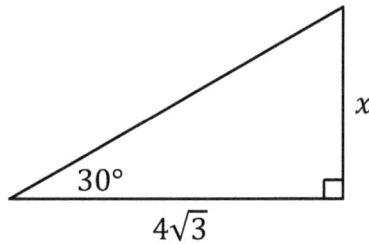

x

30°

$4\sqrt{3}$

76) Find the perimeter and area.

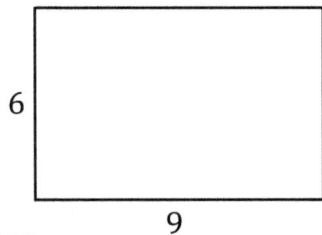

6

9

77) Find the area.

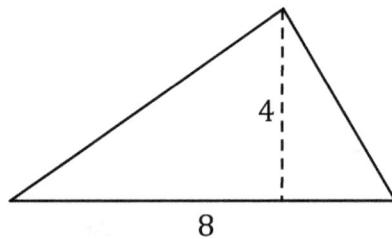

4

8

78) Find the area.

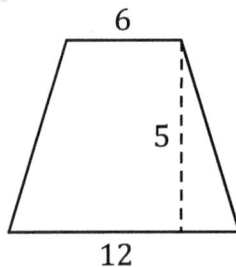

6

5

12

79) Find the area and circumference.

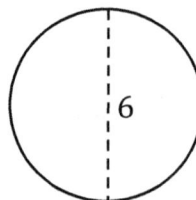

6

100

80) Determine the diameter, D.

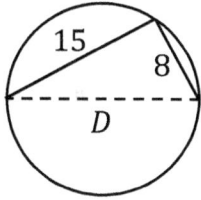

81) $R = 6$. C = center. Determine s and α.

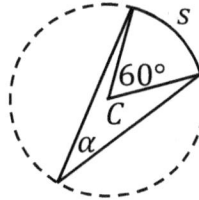

82) Find the slope-intercept equation of the line passing through $(-7, -2)$ and $(9, 6)$.

83) How many unique ways are there to arrange the letters of the word STATISTICS?

84) Three fair six-sided dice are tossed simultaneously. What is the probability that the total of all three dice will equal 10?

85) A bag contains 5 blue balls and 4 red balls. If two balls are randomly selected from the bag (without putting the first ball back into the bag), what is the probability that both balls will be the same color?

Note: Use the following data values to answer Questions 86-89:

$$3, 5, 7, 7, 8$$

86) What is the arithmetic mean?

87) What is the median?

88) What is the range?

89) What is the standard deviation?

90) Convert $120°$ to radians.

91) Convert $\frac{3\pi}{4}$ radians to degrees.

92) Evaluate $\sin 60°$, $\cos 45°$, and $\tan 30°$.

93) Evaluate $\sec 60°$, $\csc 45°$, and $\cot 30°$.

94) Evaluate $\sin 150°$, $\cos 240°$, and $\tan 180°$.

95) Find $\sin^{-1}\left(\frac{1}{2}\right)$ in Quadrant II.

96) Find $\cos^{-1}(-1)$.

97) Determine x.

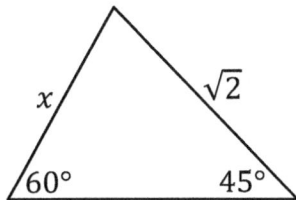

98) Determine x.

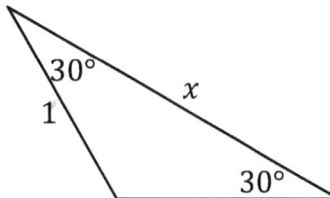

99) Solve $6 \cos \theta + 2 = 5$.

100) Solve $4 \sin \theta \cos \theta = \sqrt{3}$.

101) Solve $9 \sin^2 \theta + \cos^2 \theta = 3$.

102) Solve $\sin^2 \theta + \sin \theta - 2 = 0$.

103) Evaluate $\ln(1)$.

104) Evaluate $\log_{10}(1000)$.

105) Evaluate e^0.

106) Evaluate $\log_{10}(0.01)$.

107) Evaluate $\ln(e^2)$.

108) Evaluate $\log_2(64)$.

109) For $f(x) = x^2 - \sqrt{x}$, evaluate $f(9)$.

110) What are the domain and range for $f(x) = -\sqrt{x-2}$?

111) What are the domain and range for $f(x) = 3x^2$?

112) Evaluate $\lim\limits_{x \to 3}(x^2 - 2x + 4)$.

113) Evaluate $\lim\limits_{x \to 0^+}\left(\frac{1}{x}\right)$ and $\lim\limits_{x \to 0^-}\left(\frac{1}{x}\right)$.

114) Evaluate $\lim\limits_{x \to \infty}\left(\frac{6x+3}{2x-8}\right)$.

115) Evaluate $\lim\limits_{x \to 2}\left(\frac{x^2-4}{x-2}\right)$.

116) Evaluate $\lim\limits_{x \to 0}\left(\frac{\tan x}{x}\right)$.

117) Find $\frac{d}{dx}$ of $4x^3$.

118) Find $\frac{d}{dx}$ of x.

119) Find $\frac{d}{dx}$ of $\sqrt{x}$.

120) Find $\frac{d}{dx}$ of $(3x^2 - 5)^4$.

121) Find $\frac{d}{dx}$ of $\frac{x^2+6}{x+2}$.

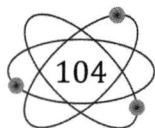

122) Find $\frac{d}{dx}$ of $4\sin^2 x$.

123) Find $\frac{d}{dx}$ of $4\sin(x^2)$.

124) Find $\frac{d}{dx}$ of $\sec x$.

125) Find $\frac{d}{dx}$ of $\cos^{-1} x$, where $|x| < 1$.

126) Find $\frac{d}{dx}$ of $x \sin x$.

127) Find $\frac{d}{dx}$ of e^{3x}.

128) Find $\frac{d}{dx}$ of $\ln(3x)$.

129) Find $\frac{d}{dx}$ of 2^x.

130) Find $\frac{d}{dx}$ of x^x.

131) Find $\frac{d}{dx}$ of $\log_{10} x$.

132) For $f(x) = x^2 - 3x + 7$, evaluate $f'(4)$.

133) Find $\frac{d^2}{dx^2}$ of $5x^4 - 4x^3 + 3x^2 - 2x + 1$.

134) Find the relative and absolute extrema of $x^5 - 15x^3$ over $[-4, 5]$.

135) What is $\int 24x^3 \, dx$?

136) What is $\int \frac{dx}{x^3}$?

137) What is $\int \frac{dx}{x}$?

138) What is $\int \sqrt{x} \, dx$?

139) Evaluate $\int_{x=0}^{\pi/2} \sin x \, dx$.

140) What is $\int \tan x \, dx$?

141) What is $\int \sec x \, dx$?

142) What is $\int e^{-ax} \, dx$?

143) What is $\int \ln x \, dx$?

144) What is $\int \sinh x \, dx$?

145) What is $\int \frac{dx}{\sqrt{1-x}}$, where $x < 1$?

146) What is $\int \frac{dx}{\sqrt{1-x^2}}$, where $|x| < 1$?

147) What is $\int x \, e^{x^2} \, dx$?

148) What is $\int xe^x \, dx$?

149) What is $\int \cos^2 x \, dx$?

150) What is $\int \cos^3 x \, dx$?

151) What is $\int \sec^3 x \, dx$?

152) What is $\int \sec^4 x \, dx$?

7 Axioms, Theorems, Laws, etc.

Note: In this chapter, we have tried to avoid repeating axioms, theorems, and principles that were included as formulas in Chapter 4.

1) What is the transitive property?

2) What is the reflexive property?

3) What is the symmetric property?

4) What is the Unique Factorization Theorem (or Fundamental Theorem of Arithmetic)?

5) What is the Prime Number Theorem?

6) What is Chebyshev's Theorem (regarding numbers)?

7) What is Bertrand's Postulate (regarding numbers)?

8) What are the Peano Axioms (also known as Peano-Dedekind)?

9) How does the discriminant of the quadratic equation relate to the roots?

10) What are the Remainder Theorem and Factor Theorem (regarding polynomials)?

11) What is Fermat's Last Theorem (regarding numbers)?

12) What are the Four Squares Theorem and Lagrange's Theorem (regarding numbers)?

13) What is the Four Color Theorem (regarding maps)?

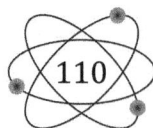

14) What is the Handshaking Lemma?

15) What are Euclid's Axioms?

16) What is the Angle Sum Theorem (for triangles)?

17) What is the Angle Sum Theorem for Polygons?

18) What is the Triangle Inequality Theorem?

19) When a transversal intersects parallel lines, which pairs of angles are congruent?

20) Which theorems can be used to demonstrate that two triangles are congruent?

21) Which theorems can be used to demonstrate that two triangles are similar?

22) What is the Perpendicular Bisector Equidistant Theorem?

23) What is the Triangle Bisector Theorem?

24) What is the Midsegment Theorem (regarding triangles)?

25) What is the Inscribed Angle Theorem?

26) What is Thales's Theorem (regarding circles)?

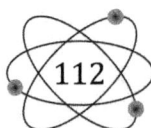

27) What is the Tangent-Chord Theorem?

28) What is Ptolemy's Theorem (regarding quadrilaterals)?

29) What is the Parallelogram Law?

30) What is Morley's Theorem (regarding triangles)?

31) What is Pappus's Centroid Theorem (for volume of revolution)?

32) Which incorrect notion is referred to as the "Law of Averages"?

33) What is the Law of Large Numbers?

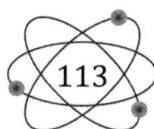

34) What is the Central Limit Theorem (regarding statistics)?

35) What is Bernoulli's Theorem (regarding probability)?

36) What is the Addition Law (regarding probability or set theory)?

37) What are the Absorption Laws (regarding set theory)?

38) What are De Morgan's Laws (regarding set theory)?

39) What is the Inclusion-Exclusion Principle (regarding set theory)?

40) What is Osborne's Rule (regarding hyperbolic functions)?

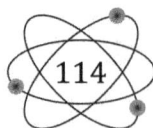

Test Your General Math Knowledge

41) What is the Fundamental Theorem of Algebra?

42) What is the Fundamental Theorem of Calculus?

43) What is the Mean Value Theorem?

44) What is the Extreme Value Theorem?

45) What is the Intermediate Value Theorem?

46) What is l'Hôpital's Rule (regarding calculus)?

47) What is Rolle's Theorem (regarding calculus)?

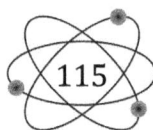

48) What is the Derivative Test (regarding calculus)?

49) What is Abel's Test (regarding series)?

50) What is the Comparison Test (regarding series)?

51) What is Dirichlet's Test (regarding series)?

52) What is the Ratio Test (regarding series)?

53) What is the Jordan Curve Theorem?

54) What is the Riemann Zeta Hypothesis?

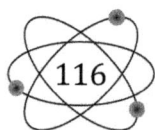

8 Math History Challenge

1) Who is the famed author of the *Elements*? When did he live?

2) Who is the most famous right triangle theorem named after? When did he live?

3) Who is the triangle of binomial coefficients named after (in the West)? When did he live?

4) Who is credited with the proof that $\sum_{n=1}^{\infty} \frac{1}{n^2} = \frac{\pi^2}{6}$? When did he live?

5) Who wrote a famous book which has the word "al-jabr" in the title? When did he live? Which subject is he considered the father of, which derives from the word "al-jabr"?

6) Who is the (x, y) coordinate system named after? When did he live?

7) Who showed that the volume of a sphere is two-thirds the volume of a cylinder of the same radius and height in his book, *Sphere and Cylinder*? When did he live?

8) Who solved a rabbit problem with the sequence 1, 1, 2, 3, 5...? (In modern times, this popular sequence begins 0, 1, 1...) When did he live?

9) Who wrote *The Nine Chapters on the Mathematical Art*, which included an algorithm to compute π and expressed answers using decimal fractions? When did he live?

10) Who used the terms "fluent" and "fluxion" in his formulation of calculus? When did he live? What did these terms mean?

11) Who developed calculus independently of the answer to #10? When did he live?

12) Who wrote a book on astronomy that included several chapters of math, worked with zero and negative numbers, and found the area of a cyclic quadrilateral? When did he live?

13) Who wrote *Disquisitiones Arithmeticae* and proved the fundamental theorem of algebra? When did he live?

14) Who is credited with five geometric theorems, including that the angle inscribed in a semicircle is a right angle? When did he live?

15) Who found extreme values, tangents, and introduced the variational principle of least time before the development of calculus? When did he live?

16) Who determined that π lies between 3.1415926 and 3.1415927, a record that held up for nearly one millennium? When did he live?

17) Who wrote *Treatise on the Circumference*, in which he determined 2π to nine places in sexagesimal (equivalent to 16 decimal places)? When did he live?

18) Who worked on algebraic and differential invariants (including work on conservation laws that bears her name), and did highly creative work on ring theory? When did she live?

19) Who is credited with first measuring the size of the earth and also measured the tilt of the earth's axis? When did he live?

20) Who is credited with introducing the concept of the logarithm? When did he live?

21) Who wrote *Ars Conjectandi*, used the term "integral" in calculus, and developed the weak law of large numbers? When did he live?

22) Who introduced l'Hôpital's rule to his student, l'Hôpital, and posed the brachistochrone problem? When did he live?

23) Who wrote the eight-volume work, *The Conics*? When did he live?

24) Who is famous for his incompleteness theorems? When did he live?

25) Who wrote the *Almagest* and made detailed tables of chords? When did he live?

26) Who made a table of chords prior to the answer to #25 and is credited with discovering the precession of the equinoxes? When did he live?

27) Who was a self-taught mathematician who worked on number theory, elliptic functions, continued fractions, and partitions (including the circle method)? When did he live?

28) Who wrote *Ars Magna*, a comprehensive treatise on algebra? When did he live?

29) Who wrote *l'Algebra* and addressed the issue of complex numbers? When did he live?

30) Who received the Abel Prize for his work on geometry, the Wolf Prize for his work on group theory, geometry, and topology, and the Nemmers Prize? When was he born?

31) Who is credited with discovering the triangle of binomial coefficients in China centuries before it was discovered in the West? When did he live?

32) Who is known for presenting #31's triangle and also for his work on magic squares and magic circles? When did he live?

33) Who presented #31's triangle of binomial coefficients again in his work on algebraic computations and dimensions of geometric figures? When did he live?

34) Which Persian mathematician worked on cubic equations in the context of intersecting conic sections, worked on higher roots of whole numbers, and also described a triangle of binomial coefficients? When did he live?

35) Who was the first female mathematician to receive the prestigious Fields Medal for her work on the symmetry of curved surfaces? When was she born?

36) Who is credited for founding set theory? When did he live?

37) Who declined the prestigious Fields Medal that he was awarded in 2006 for his work on the Poincaré conjecture? When was he born?

38) Who derived an equation that is fundamental to the calculus of variations, was a key figure in the formulation of mechanics, and proved that any positive integer can be written as the sum of at most four squares? When did he live?

39) Which mathematician inspired the movie *A Beautiful Mind*? When was he born?

40) Which female mathematician used the pseudonym M. Le Blanc as a means to obtain lecture notes and correspond with famous mathematicians? When did she live?

41) Who is credited with proving Fermat's Last Theorem? When was he born? Who helped when holes were pointed out in his initial proof?

42) Who wrote *Arithmetica Infinitorum*, which #10 built upon to develop calculus? When did he live?

43) Who are the four mathematicians who won the Fields Medal, the Abel Prize, and the Wolf Prize?

44) Who proved that the distance traveled during uniform acceleration is the product of the average speed and the elapsed time? When did he live? What is the theorem called?

45) Who was highly influential with his work on algebraic geometry and received the Fields Medal for it in 1966? When was he born?

46) Who developed a method of indivisibles similar to integral calculus? When did he live?

47) Which female mathematician received the Abel Prize in 2019? When was she born?

48) Which mathematicians received the Fields Medal in 2018?

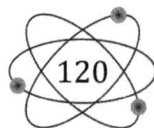

9 Math Abbreviation Challenge

1) What does No. stand for (as in No. 32)?

2) What does c. stand for (as in c. 400 BC)?

3) What do LCD and LCM stand for (in arithmetic)?

4) What do GCF, HCF, and GCD stand for (in arithmetic)?

5) What does PEMDAS stand for (in arithmetic)?

6) What does TI stand for (in technology)?

7) What does CPA stand for (in accounting)?

8) What do AR and AP stand for (in accounting)?

9) What do CR and DR stand for (in accounting)?

10) What does ROI stand for (in accounting)?

11) What does IRA stand for (in accounting)?

12) What does P&L stand for (in accounting)?

13) What does LTL stand for (in accounting)?

14) What does IRS stand for (in accounting)?

15) What does STEM stand for (in education)?

16) What do M.S. and Ph.D. stand for (in education)?

17) What does APR stand for (in finance)?

18) What does UPC stand for (in finance)?

19) What does RPI stand for (in finance)?

20) What does EMV stand for (in finance)?

21) What does ARM stand for (in finance)?

22) What does CD stand for (in finance)?

23) What does NYSE stand for (in finance)?

24) What does GNP stand for (in finance)?

25) What does EFT stand for (in finance)?

26) What does YTD stand for (in finance)?

27) What does NASDAQ stand for (in finance)?

28) What does ASCII stand for (in computing)?

29) What does IMO stand for (in math competitions)?

30) What does AMC stand for (in math competitions)?

31) What does foil stand for (in algebra)?

32) What does QE stand for (in algebra)?

33) What does FLT stand for (in number theory)?

34) What do SSS and AA stand for (in geometry)?

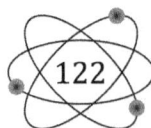

35) What do SAS, ASA, and AAS stand for (in geometry)?

36) What does CPCTC stand for (in geometry)?

37) What does PT stand for (in geometry)?

38) What does PBET stand for (in geometry)?

39) What do i.e. and e.g. stand for (in writing)?

40) What does et al. stand for (in writing)?

41) What does n.b. stand for (in writing)?

42) What does viz. stand for (in writing)?

43) What do vs. and v.s. stand for (in writing)?

44) What does cf. stand for (in writing)?

45) What does wlog stand for (in writing)?

46) What does wrt stand for (in proofs)?

47) What does iff stand for (in proofs)?

48) What does STP stand for (in proofs)?

49) What do QED and QEF stand for (in proofs)?

50) What does SI stand for (regarding units)?

51) What does lb stand for (regarding units)?

52) What does oz stand for (regarding units)?

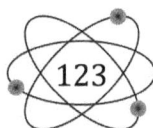

53) What do qt, pt, and c stand for (regarding units)?

54) What do tbsp and tsp stand for (regarding units)?

55) What do cc and L stand for (regarding units)?

56) What do F, C, and K stand for (regarding units)?

57) What does mph stand for (regarding units)?

58) What does rpm stand for (regarding units)?

59) What do c, d, da, f, G, h, k, m, μ, n, p, and T stand for (regarding prefixes)?

60) What does AMS stand for (regarding professional associations)?

61) What does MAA stand for (regarding professional associations)?

62) What does EMS stand for (regarding professional associations)?

63) What does JAMS stand for (regarding journals)?

64) What does EJM stand for (regarding journals)?

65) What does IMA stand for (regarding journals)?

66) What does sohcahtoa stand for (in trigonometry)?

67) What does ASTC stand for (in trigonometry)?

68) What does SHM stand for (in trigonometry or in physics)?

69) What does exp stand for (in precalculus)?

70) What do log and ln stand for (in precalculus)?

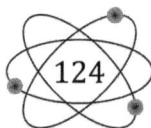

71) What does DNE stand for (regarding limits)?

72) What does RSA stand for (in cryptography)?

73) What does rv stand for (in prob-stat)?

74) What do sd and se stand for (in prob-stat)?

75) What does rms stand for (in prob-stat)?

76) What does IQR stand for (in prob-stat)?

77) What do p.d.f. and p.g.f. stand for (in prob-stat)?

78) What does c.d.f. stand for (in prob-stat)?

79) What does EDA stand for (in prob-stat)?

80) What does i.i.d. stand for (in prob-stat)?

81) What does erf stand for (in prob-stat)?

82) What does pmcc stand for (in prob-stat)?

83) What does FTOC stand for (in calculus)?

84) What does MVT stand for (in calculus)?

85) What does sgn stand for (in calculus)? (It's not sign.)

86) What do DE, ODE, and PDE stand for (in advanced math)?

87) What does BC stand for (in advanced math)?

88) What does RKM stand for (in advanced math)?

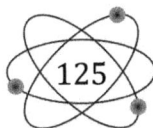

89) What does TSP stand for (in advanced math)?

90) What does gf stand for (in advanced math)?

91) What do RE and IM stand for (in advanced math)?

92) What does cc stand for (in advanced math)?

93) What does adj. stand for (in advanced math)? (Not the trigonometry abbreviation.)

94) What does tr stand for (in advanced math)?

95) What does dof stand for (in advanced math)?

96) What does FFT stand for (in advanced math)?

97) What do SU and SO stand for (in advanced math)?

98) What does SUSY stand for (in advanced math or physics)?

99) What does TOE stand for (in advanced math or physics)?

100) What does GUT stand for (in advanced math or physics)?

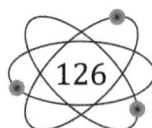

Answer Key

1 Test Your Vocabulary

1) borrow
2) carry over
3) ordinal
4) prime
5) composite
6) round
7) compass
8) protractor
9) abacus
10) slide rule
11) ruler
12) straight edge
13) French curve
14) Fibonacci
15) Lucas
16) infinity
17) infinitesimal
18) finite
19) zero, nought
20) one, unity
21) numerals
22) enumerate
23) gross, net
24) quantitative, qualitative
25) even, odd
26) parity
27) pair, triple, quadruple

28) equate
29) equivalent
30) exactly, approximately
31) estimated, rounding
32) truncate
33) operations
34) operators
35) factorials
36) equal, inequal
37) approximately equal to, on the order of
38) less than, greater than
39) less than or equal to, greater than or equal to
40) addend + addend = sum
41) minuend – subtrahend = difference
42) factor × factor = product
43) dividend ÷ divisor = quotient
44) multiples
45) remainder
46) squared, cubed
47) square root
48) cube root
49) fractions
50) proper fraction, improper fraction

51) reduced
52) simple fraction, compound fraction
53) mixed number
54) numerator, denominator
55) decimal
56) repeating (or recurring) decimal
57) scientific notation
58) percent, percentage
59) ratio
60) proportion
61) positive, negative
62) nonzero
63) nonnegative
64) sign
65) integer
66) whole
67) natural
68) rational, irrational
69) real, imaginary
70) complex
71) decimal point
72) digits
73) place value
74) units digit, thousandths place
75) decimal places

76) significant figures

77) trailing zeroes (or zeros)

78) parentheses (note that parenthesis would be singular)

79) brackets

80) half

81) third

82) fourth, quarter

83) twice, double

84) million, billion, trillion

85) hundredth, thousandth

86) millionth, billionth

87) dozen, gross

88) googol, googolplex

89) decimal, binary

90) hexadecimal, octal

91) sexagesimal

92) absolute value

93) base

94) raised, power

95) exponent

96) reciprocal, multiplicative inverse

97) conversion

98) rate

99) displacement

100) speed

101) distance, time

102) velocity, speed

103) acceleration

104) uniform

105) bit

106) byte

107) digital

108) analog

109) common denominator

110) factor tree

111) greatest common factor (or divisor)

112) decomposition (since prime factorization is two words; but we'll award partial credit for factorization)

113) divisible

114) perfect squares

115) rationalizing

116) value

117) revenue, cost

118) principal, interest

119) simple interest, compound interest

120) discount

121) Pythagorean triples

122) perfect

123) Euclid

124) triangular

125) amicable

126) twin primes

127) relatively prime (we'll accept coprime, even though it is a single word)

128) Mersenne primes

129) magic

130) Latin

131) Graeco-Latin

132) mantissa

133) matrix, determinant

134) algorithm

135) iteration

136) lever, fulcrum (or hinge)

137) Hero

138) commutative

139) associative

140) distributive

141) inverse

142) transitive

143) reflexive

144) graph

145) plot

146) Venn

147) tree

148) tessellation

149) pie

150) bar

151) histogram

152) box plot, whiskers

153) stem, leaf

154) scatter

155) axis, horizontal, axis, vertical

156) scale

157) coordinates, point

158) Cartesian coordinates

159) ordered

160) origin

161) axis, abscissa

162) axis, ordinate

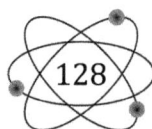

163) axes, Quadrants

164) semilog

165) intercept

166) straight line

167) slope

168) rise, run

169) interpolate

170) extrapolate

171) circle

172) radius

173) diameter

174) tangent

175) secant

176) chord

177) arc

178) sector, segment

179) ellipse

180) parabola

181) hyperbola

182) foci (or focuses)

183) major

184) branches

185) directrix

186) eccentricities

187) asymptotes

188) conic sections, cone

189) cycle

190) epicycloid

191) deferent

192) symmetry

193) bilateral symmetry

194) region

195) simple

196) knot

197) exponential growth

198) exponential decay

199) random walk

200) osculate

201) mensuration

202) perspective, vanishing

203) line, line segment

204) ray

205) parallel

206) intersect

207) askew (or skew)

208) perpendicular

209) normal, orthogonal

210) cusp

211) angle

212) vertex, node

213) arm

214) degrees

215) arc minute

216) arc second

217) radian

218) grade

219) bearing

220) polygon

221) sides

222) triangle, quadrilateral, pentagon, hexagon

223) heptagon, octagon, nonagon (or enneagon), decagon

224) hendecagon, dodecagon

225) square, rectangle

226) parallelogram, rhombus

227) trapezoid (or trapezium)

228) kite

229) pentagram

230) hexagram

231) right

232) acute

233) obtuse

234) reflexive angle

235) entrant (the re- is already there)

236) hypotenuse

237) legs

238) base, altitude, height

239) apex

240) bisect

241) trisect

242) midpoint

243) median

244) diagonal

245) perimeter

246) area

247) volume

248) complementary, supplementary

249) vertical

250) interior

251) exterior

252) plane

253) golden

254) semicircles

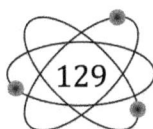

255) circumference

256) pi

257) arc length

258) central

259) inscribed

260) subtends

261) equilateral

262) equiangular

263) regular

264) isosceles

265) scalene

266) equidistant

267) locus

268) clockwise, counterclockwise

269) annulus

270) solid

271) surface

272) cube

273) cuboid

274) pyramid

275) prism

276) truncated cube

277) sphere

278) hemispheres

279) ball

280) spherical shell

281) latitude

282) longitude

283) poles

284) great circle

285) meridian

286) equator

287) triangulation

288) cylinder

289) cone

290) slant height

291) frustum

292) developable surface

293) ruled surface

294) disc

295) ring

296) hoop

297) torus

298) spiral

299) helix

300) collinear

301) coplanar

302) coincident

303) concurrent

304) concentric

305) coaxial

306) center of mass (or center of gravity)

307) centroid

308) geodesic

309) octants

310) polyhedron

311) faces

312) Platonic solids

313) Archimedean solids

314) tetrahedron

315) octahedron

316) dodecahedron

317) icosahedron

318) icosidodecahedron

319) truncated tetrahedron

320) rhombohedron

321) Möbius strip (or band)

322) Klein bottle

323) oblong

324) apothem

325) lateral surface (or face)

326) dimension

327) fractal

328) hyperplane

329) tesseract, hypercube

330) glome, hypersphere

331) polytope

332) simplex

333) Schläfli symbol

334) arithmetic

335) finance

336) economics

337) algebra

338) abstract algebra

339) linear algebra

340) geometry

341) cryptography

342) probability

343) combinatorics

344) statistics

345) applied

346) topology

347) number

348) decision

349) game

350) heuristics

351) trigonometry

352) chaos

353) symbolic logic

354) Boolean logic

355) set

356) group

357) calculus

358) differential equations

359) graph

360) numerical analysis (or methods)

361) biometry

362) axiom, postulate

363) first principles

364) theorem

365) proof

366) law

367) direct

368) indirect

369) synthetic, analytic

370) proposition

371) conjecture

372) lemma, theorem

373) corollary

374) generalized

375) deduction

376) induction

377) inference

378) truth table

379) liar paradox

380) fallacy

381) invalid

382) conditional

383) antecedent, consequent

384) converse

385) inverse

386) contrapositive

387) negation

388) implies

389) if and only if

390) necessary

391) necessary, sufficient

392) analogy

393) a priori

394) a posteriori

395) circular argument (or reasoning)

396) contradiction

397) paradox

398) contingent

399) counterexample

400) causation

401) compound

402) conjunction

403) inclusive disjunction

404) exclusive disjunction

405) mutually exclusive

406) logically equivalent

407) equipollent (or equipotent)

408) tautology

409) self-reference

410) syllogism

411) dichotomy

412) quod erat demonstrandum

413) quod erat faciendum

414) reductio ad absurdum

415) centesimal

416) chromatic

417) rectilinear

418) variable, unknown

419) constant

420) coefficient

421) parameter

422) equation, expression

423) terms

424) inequality, interval

425) given

426) solve, simplify

427) combine like terms

428) isolate the unknown

429) solutions

430) trivial solution

431) undefined (division by zero)

432) indeterminate (any value of x works)

433) no solution (no value of x works)

434) all real numbers (and even complex ones)

435) insoluble (only an imaginary number works)

436) incompatible, inconsistent

437) under-determined

438) transcendental

439) formulas

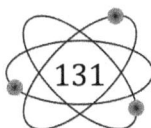

440) evaluate

441) plug and chug

442) distribute

443) factor out

444) expanded

445) linear

446) polynomials

447) quadratic

448) discriminant

449) degree

450) cubic, quartic, quintic

451) cancels

452) reducible

453) cross-multiply

454) simultaneous

455) eliminated

456) substitution

457) completing, square

458) binomials

459) Pascal's triangle

460) trinomials

461) directly

462) linearly proportional

463) inversely proportional

464) identities

465) synthetic division

466) homogeneous, fourth

467) open interval, closed interval

468) unconditional (or absolute) inequality

469) canonical

470) conjugate surds

471) unique

472) distinct

473) degenerate

474) extraneous, superfluous

475) redundant

476) dependent

477) Diophantine

478) possible

479) congruent

480) similar

481) convex

482) concave

483) stellated

484) transversal

485) corresponding

486) alternate interior

487) alternate exterior

488) inscribed, circumscribed

489) closed

490) parallelepiped

491) cuboctahedron

492) construct

493) elevation (or inclination)

494) depression (or declination)

495) conjugate

496) inscribed

497) confocal

498) concyclic

499) incidence

500) oblique

501) dihedral

502) genus

503) Fermat

504) projection

505) section (such as a cross-section)

506) zone

507) radical

508) probability, likelihood

509) odds

510) empirical

511) random

512) permutation

513) combination

514) data

515) measurement

516) raw

517) encrypt

518) table

519) population, sample

520) representative

521) trial

522) replicated

523) outlier, anomaly

524) arithmetic mean

525) geometric mean

526) harmonic mean

527) weighted average (or mean)

528) median

529) frequency

530) relative frequency

531) mode

532) bimodal

533) categorical, nominal

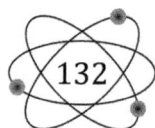

534) law of averages

535) independent (or explanatory), dependent

536) error

537) random

538) systematic

539) noise

540) relative

541) percent

542) residual

543) quartiles

544) percentile

545) range

546) interquartile range

547) deviation

548) expected value

549) variance

550) standard deviation

551) spread, dispersion

552) robust

553) trial and error

554) fit

555) least squares

556) linear regression

557) chi-squared

558) confidence level

559) reliability

560) correlation

561) spurious

562) confounding

563) lurking

564) control

565) skewness

566) bias

567) accuracy

568) precision

569) degrees of freedom

570) Gaussian distribution (normal distribution)

571) standardized

572) model

573) simulation

574) stochastic

575) set

576) group

577) element, member

578) union

579) intersection

580) empty, null

581) identity element, neutral element

582) inverse element

583) reflection

584) translation

585) rotation

586) shear (since all rectangles are parallelograms, the "non-right" indicates that it is no longer equiangular)

587) equivalence

588) cardinal

589) disjoint

590) cyclic

591) family

592) denumerable

593) mapping

594) pairwise

595) difference

596) fuzzy logic

597) singleton

598) adjacent, opposite, hypotenuse

599) sine (opposite over hypotenuse)

600) cosine (adjacent over hypotenuse)

601) tangent (opposite over adjacent)

602) cosecant (reciprocal of sine, hypotenuse over opposite)

603) secant (reciprocal of cosine, hypotenuse over adjacent)
Note: the co's don't go together (secant is the reciprocal of cosine, whereas cosecant is the reciprocal of sine)

604) cotangent

605) arcsine (we'll accept arcsin)

606) principal

607) sinusoidal

608) unit circle

609) parametric

610) vector, scalar

611) antiparallel

612) components

613) unit vector

614) taxicab norm

615) resolved

616) resultant

617) displacement

618) amplitude

619) cycle

620) revolution (not rotation; rotation refers to traveling in a circle, whereas one revolution means one cycle)

621) period

622) frequency

623) wavelength

624) angular frequency

625) phase

626) oscillation

627) harmonic

628) periodic

629) function

630) domain

631) range

632) even

633) odd

634) concavity

635) concave up, concave down

636) multiplicatively separable, additively separable

637) bounded

638) unbounded

639) discrete

640) continuous

641) discontinuity

642) singularity

643) root, zero

644) step function

645) signum function

646) arguments

647) monotonic

648) strictly

649) bilinear

650) exponential function

651) logarithm

652) half-life

653) hyperbolic functions

654) involution

655) complex plane

656) complex conjugate

657) modulus

658) phase angle

659) quaternion

660) Gaussian

661) limit

662) vanish

663) instantaneous

664) increment

665) neighborhood

666) superstitious

667) jump

668) ill-conditioned

669) partition

670) derivative

671) second

672) higher-order

673) differentiate

674) differentiable

675) analytic, holomorphic

676) smooth

677) indefinite integral, definite integral

678) anti-derivative

679) integration

680) integrand

681) arbitrary constant

682) dummy

683) Riemann

684) chain

685) related rates

686) critical, stationary

687) derivative

688) relative

689) absolute

690) inflection

691) turning

692) optimization

693) constraint

694) integration by parts

695) partial fractions

696) implicit differentiation

697) improper integrals

698) partial derivative

699) del

700) gradient

701) divergence

702) curl

703) Laplacian

704) curvature

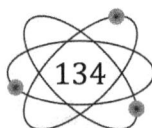

705) multivariate

706) sequence, series

707) progression

708) index

709) alternating

710) arithmetic

711) geometric

712) harmonic

713) power

714) asymptotic

715) exponential

716) Taylor

717) Maclaurin

718) Fourier

719) diverges, converges

720) ratio

721) comparison

722) binomial expansion (or series)

723) tensor

724) summation

725) Kronecker delta

726) Levi-Civita

727) product

728) invariant

729) polar

730) spherical (or spherical polar)

731) cylindrical (or cylindrical polar)

732) radial

733) azimuthal

734) direction cosines

735) axial planes

736) cycloid

737) hypocycloid

738) astroid (it has no "e")

739) cardioid

740) catenary

741) brachistochrone

742) tautochrone (or isochrone)

743) quadric surface

744) latus rectum

745) antipodal

746) spheroid

747) prolate, oblate

748) ellipsoid

749) paraboloid

750) hyperbolic, saddle point

751) elliptic

752) hyperboloid, sheets

753) Gabriel's horn (or Torricelli's trumpet)

754) solid

755) steradian

756) linearly independent

757) rank

758) identity

759) square

760) diagonal

761) transpose

762) symmetric

763) antisymmetric (or skew-symmetric)

764) inverse

765) cofactor

766) trace

767) Hermitian, self-adjoint

768) orthogonal

769) augmented

770) echelon form

771) Gaussian elimination

772) characteristic polynomial (for this to be the characteristic equation, you would also need to set it equal to zero)

773) eigenvector, eigenvalue

774) singular

775) conformable

776) adjoint

777) block diagonal

778) banded

779) inner, outer

780) Jacobian

781) ordinary differential

782) order

783) partial differential

784) boundary conditions

785) boundary value

786) Dirichlet, Neumann (aka first-type, second-type)

787) linear first-order

788) integrating factor

789) linear, constant coefficients

790) separable first-order

791) homogeneous first-order

792) autonomous

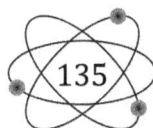

793) general, complete

794) particular

795) auxiliary (or characteristic)

796) exact differential

797) coupled

798) recurrence, difference

799) damped

800) critically damped

801) stability

802) Euler

803) Runge-Kutta

804) Newton's (or Newton-Raphson)

805) bisection

806) quadrature

807) Simpson

808) trapezoidal

809) Legendre polynomials

810) Bessel functions

811) gamma function

812) beta function

813) zeta function

814) elliptic

815) Dirac delta function

816) Fourier transform

817) Laplace transform

818) sources, sinks

819) perturbation

820) morphism

821) Monte Carlo

822) manifold

823) metric tensor

824) Lorentz transformation

825) Lagrangian

826) action

827) Lagrange multipliers

828) Hamiltonian

829) contour integral

830) residue

831) Hilbert

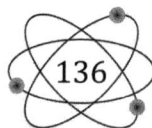

136

2 Number Challenge

1) 3

2) 0

3) 1

4) 3.142 (precisely, 3.14159265...)

5) 1,000,000

6) 256

7) ±3

8) 4 (bonus for the complex roots $-2 \pm 2i\sqrt{3}$)

9) 1.414

10) 1.732

11) 2.718

12) 0.577

13) 1.618

14) 1830 (find pairs that add up to 60)

15) 62.5%

16) $0.\overline{36}$

17) 11110

18) 45

19) 64

20) 125

21) 10

22) 100,000

23) 1000

24) 86,400

25) 5280

26) 36

27) 9 (although 3 feet = 1 yard, square feet is different, as shown visually below)

28) 2.54

29) 660

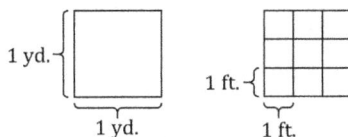

1 yd. — [square] 1 yd.   1 ft. — [grid] 1 ft.

30) 1.15

31) 2000

32) 2240

33) 2205

34) 4840

35) XCIV

36) 945

37) 6, 12, 8

38) 4, 6, 4

39) 12, 30, 20

40) 90°

41) 360°

42) 42 (bonus for the question)

43) $1 : 1 : \sqrt{2}$

44) $1 : \sqrt{3} : 2$

45) 60°

46) 108°

47) 109.5°

48) 70.5°

49) 12

50) 2π

51) 4π

52) 57.3°

53) 1

54) 1, 1, 120

55) 1, 7, 21, 35, 35, 21, 7, 1

56) $0, \frac{1}{2}, \frac{\sqrt{2}}{2}$ (same as $\frac{1}{\sqrt{2}}$), $\frac{\sqrt{3}}{2}, 1$

57) $1, \frac{\sqrt{3}}{2}, \frac{\sqrt{2}}{2}$ (same as $\frac{1}{\sqrt{2}}$), $\frac{1}{2}, 0$

58) $0, \frac{\sqrt{3}}{3}$ (same as $\frac{1}{\sqrt{3}}$), $1, \sqrt{3},$ undefined

59) $-\frac{1}{2}, \frac{\sqrt{2}}{2}$ (same as $\frac{1}{\sqrt{2}}$), $-\sqrt{3}$

60) $2, -\frac{2\sqrt{3}}{3}$ (same as $-\frac{2}{\sqrt{3}}$), $-\sqrt{3}$

61) $\frac{\sqrt{6}-\sqrt{2}}{4}$, which equals $\frac{\sqrt{2}}{4}\left(\sqrt{3}-1\right)$

62) $\sin^{-1}\left(\frac{1}{2}\right) = 30°$ or $150°$, $\cos^{-1}\left(-\frac{\sqrt{2}}{2}\right) = 135°$ or $225°$, $\tan^{-1}(-1) = 135°$ or $315°$

63) 2, 0, and 0.693

64) 0

65) $\frac{\sqrt{2}}{2} + i\frac{\sqrt{2}}{2}$ or $-\frac{\sqrt{2}}{2} - i\frac{\sqrt{2}}{2}$; these can be written as $\pm\frac{\sqrt{2}}{2}(1+i)$

66) 26, 10, 11 (of spacetime; subtract 1 if you just want space)

67) 8, 24, 32, 16

68) 3/2 (not –4/27; read carefully)

69) $\sqrt{\pi}$

3 Symbol Challenge

1) +

2) the – symbol or use parentheses like ($3.50)

3) 3×4, 3·4, or (3)(4)

4) beside one another without a symbol like $5xy$

5) 8÷2, 8/2, or $2\overline{)8}$

6) slash like 2/3 or horizontal line like $\frac{2}{3}$

7) : like 4:5

8) $A{:}B = C{:}D$ or $A{:}B{::}C{:}D$

9) %

10) point (3.14) in the US, middle dot (3·14) in the UK, comma (3,14) in continental Europe

11) comma (1,234) in the US, point (1.234) in Spain, underline (1_234) maritime, space (1 234) in Australia, and in rare usage even an apostrophe (1'234)

12) =

13) >, <, ≠, ≥, ≤

14) ≈

15) ~

16) ∝

17) $\sqrt{9}$ or $9^{1/2}$

18) (), [], {}, <>

19) vertical lines like $|x|$

20) …

21) I, V, X, L, C, D, M

22) ℝ or **R**

23) ℂ or **C**

24) $

25) 25¢ or $0.25

26) £

27) €

28) ¥

29) ₹

30) #

31) ±

32) ∞

33) A, B, C, D, E, F

34) π

35) θ

36) φ, ϕ

37) Δ

38) δ

39) Σ

40) Π

41) ϵ, ε

42) ω

43) μ

44) χ

45) e

46) γ

47) °

48) ', "

49) AB

50) $\overline{AB}$

51) $\overrightarrow{AB}$

52) $\overleftrightarrow{AB}$

53) ∠

54) ∟

55) $m\angle$

56) Δ

57) ⊿

58) ∥

59) ⊥

60) ∦

61) ≅

62) ~

63) m, b

64) L, W, H, D (sometimes lowercase)

65) b, h (sometimes uppercase)

66) A, P, V

67) R, D, C (sometimes r and d)

68) k

69) s

70) f

71) a, b

72) e

73) □ or ■

74) ≡, : =, $\stackrel{\text{def}}{=}$, or even ≜

75) ⇒, ⊃, or →

76) ⇔ or ↔

77) ∴

78) ∵

79) : or |

80) ∃

81) $\forall$

82) $\in$

83) $\cup, \cap$

84) $\wedge, \vee$

85) $\neg$

86) $\subset$ or $\supset$, $\subseteq$ or $\supseteq$

87) $\emptyset$

88) an overbar like $\bar{x}$

89) !

90) $\binom{n}{m}$

91) H, T

92) T, F (or lowercase)

93) ( )

94) | |

95) $x, y, 1$ (because it's a **unit** circle)

96) boldface or an arrow above like $\vec{V}$

97) a caret (or hat) above like $\hat{u}$

98) $\hat{i}, \hat{j}, \hat{k}$ or $\hat{x}, \hat{y}, \hat{z}$, or even $\hat{x}_1, \hat{x}_2, \hat{x}_3$

99) omit the arrow (or boldface) like V or use double lines like $\|\vec{V}\|$

100) subscripts i and f like x_i and x_f, or subscript nought (0) for initial and no subscript for final like x_0 and x

101) d (or x), r (or v), t

102) v (since speed is the magnitude of velocity)

103) $\vec{r}, \Delta\vec{r}$ (or $\vec{s}$ or even $\overrightarrow{AB}$)

104) $\vec{v}, \vec{a}$ (or v_x and a_x for 1D motion)

105) $\vec{\omega}, \vec{\alpha}$

106) f, T (or p or $\frac{2\pi}{\omega}$)

107) λ

108) ', "

109) m, kg, s

110) L, M, T (in dimensional analysis)

111) m/s, m/s^2

112) Hz, s

113) kg·m^2

114) N, J, W

115) kg·m/s, kg·m^2/s

116) N/C (or V/m), T

117) J, K

118) kg/m^3

119) J/K

120) A, Ω, V

121) $\vec{A} \cdot \vec{B}, \vec{A} \times \vec{B}$

122) $\ln x$

123) $Re(z), Im(z)$

124) $*$ (or an overbar)

125) x, y, z

126) r, θ

127) r, θ, φ

128) ρ (or r), θ, z

129) $\lim\limits_{x \to c} f(x)$

130) $\to$

131) $\lim\limits_{x \to c^-} f(x)$, $\lim\limits_{x \to c^+} f(x)$ (or use < or > instead of signs)

132) <<, >>

133) $\frac{dy}{dx}, y'$ (or even $\dot{y}$)

134) $\int$

135) $\frac{\partial f}{\partial x}$

136) $\nabla, \nabla \cdot, \nabla \times$

137) ∇^2 (or $\nabla \cdot \nabla$), $\square$

138) $\otimes, \odot$

139) -1 like x^{-1}

140) T like A^T

141) $\langle A | B \rangle$

142) $\dagger$

139

4 Formula Challenge

1) $a + 0 = 0 + a = a$

2) $a \times 1 = 1 \times a = a$

3) $a + b = b + a$

4) $a \times b = b \times a$

5) $(a + b) + c = a + (b + c)$

6) $(a \times b) \times c = a \times (b \times c)$

7) $a \times (b + c) = a \times b + a \times c$

8) $a + (-a) = (-a) + a = 0$

9) $a \times a^{-1} = a^{-1} \times a = 1$ where $a^{-1} = \frac{1}{a}$ (and $a \neq 0$)

10) $P\left(1 + \frac{ni}{100}\right)$

11) $P\left(1 + \frac{i}{100}\right)^n$

12) $(w + x)(y + z) = wy + wz + xy + xz$

13) $x^2 - y^2 = (x + y)(x - y)$

14) $x = \frac{-b \pm \sqrt{b^2 - 4ac}}{2a}$

15) $x^m x^n = x^{m+n}$

16) $\frac{x^m}{x^n} = x^{m-n}$

17) $x_i = \frac{\det(A_i)}{\det(A)}$, where A is the matrix of coefficients (of the variables in order) and A_i is the same as A except for replacing the ith column with the constants from the right-hand side (of the system).

18) $x^2 = ny^2 + 1$

19) $a^2 + b^2 = c^2$

20) $d = \sqrt{(x_2 - x_1)^2 + (y_2 - y_1)^2 + (z_2 - z_1)^2}$

21) $(x_m, y_m, z_m) = \left(\frac{x_1 + x_2}{2}, \frac{y_1 + y_2}{2}, \frac{z_1 + z_2}{2}\right)$

22) $(x_p, y_p, z_p) = \left(\frac{mx_2 + nx_1}{m+n}, \frac{my_2 + ny_1}{m+n}, \frac{mz_2 + nz_1}{m+n}\right)$

23) $m = \frac{y_2 - y_1}{x_2 - x_1}$

24) $y = mx + b$

25) $y - y_1 = m(x - x_1)$

26) $\frac{x}{a} + \frac{y}{b} = 1$

27) $m_1 m_2 = -1$ (equivalent to $m_2 = -\frac{1}{m_1}$)

28) $r = \frac{d}{t}$ (equivalent to $d = rt$)

29) $P = 2L + 2W, A = LW$

30) $D = 2R$ (equivalent to $R = \frac{D}{2}$)

31) $C = 2\pi R$ (or $C = \pi D$), $A = \pi R^2$

32) $s = R\theta$

33) $A = \frac{1}{2}bh$

34) $A = bh$

35) $A = \frac{d_1 d_2}{2}$

36) $A = \frac{(a+b)h}{2}$

37) $A = \frac{3L^2 \sqrt{3}}{2}$

38) $A = 2L^2\left(1 + \sqrt{2}\right)$

39) $A = \pi ab$

40) $x^2 + y^2 = R^2$

41) $y = ax^2 + b$

42) $\frac{x^2}{a^2} - \frac{y^2}{b^2} = 1$

43) $\frac{x^2}{a^2} + \frac{y^2}{b^2} = 1$

44) $y = A\sin(kx + \varphi)$

45) $r = a\theta$

46) $r = ae^{b\theta}$

47) $r^2 = a\theta, r = \frac{a}{\theta}$ (equivalent to $r\theta = a$)

48) $\frac{x+1}{x} = \frac{x}{1}$ (equivalent to $\frac{x+1}{x} = x$)

49) $(x_c, y_c, z_c) = \left(\frac{x_1 + x_2 + x_3}{3}, \frac{y_1 + y_2 + y_3}{3}, \frac{z_1 + z_2 + z_3}{3}\right)$

50) $\frac{AR}{BR} \frac{BQ}{CQ} \frac{CP}{AP} = 1$

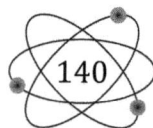

51) $AP \cdot CP = BP \cdot DP = R^2 - x^2$ where AC and BD intersect at P inside of the circle, a distance x from the center

52) $AP \cdot CP = BP \cdot DP = x^2 - R^2$ where AP and BP intersect at P outside of the circle, a distance x from the center

53) $S = 6L^2, V = L^3$

54) $S = 2(LW + LH + WH), V = LWH$

55) $S - 2(LW \sin\gamma + 2LH \sin\alpha + 2WH \sin\beta)$,

$V = LWH\sqrt{1 + 2\cos\alpha\cos\beta\cos\gamma - \cos^2\alpha - \cos^2\beta - \cos^2\gamma}$

56) $S = 4\pi R^2, V = \frac{4}{3}\pi R^3$

57) $S = 2\pi(RH + R^2), V = \pi R^2 H$

58) $S = \pi(RH + R^2), V = \frac{1}{3}\pi R^2 H$

59) $S = L^2\sqrt{3}, V = \frac{L^3\sqrt{2}}{12}$ (equivalent to $V = \frac{L^3}{6\sqrt{2}}$)

60) $S = 4\pi^2 R_1 R_2, V = 2\pi^2 R_1 R_2^2$

61) one form is $ax + by + cz = d$

62) one form is $a_1 x + b_1 y + c_1 z = d_1$ and

$a_2 x + b_2 y + c_2 z = d_2$

63) $x = R\cos t, y = R\sin t, z = kt$

64) $x^2 + y^2 + z^2 = R^2$

65) $\frac{x^2}{a^2} + \frac{y^2}{b^2} + \frac{z^2}{c^2} = 1$

66) $f + v - e = 2$

67) $\begin{vmatrix} a & b \\ c & d \end{vmatrix} = ad - bc$

68) $\begin{vmatrix} a & b & c \\ d & e & f \\ g & h & i \end{vmatrix} = aei + bfg + cdh - ceg -$

$afh - bdi$

69) $x_{ave} = \frac{x_1 + x_2 + \cdots + x_N}{N}$

70) $x_{geom} = \sqrt[n]{x_1 x_2 \dots x_N}$

71) $x_W = \frac{w_1 x_1 + w_2 x_2 + \cdots + w_N x_N}{w_1 + w_2 + \cdots + w_N}$

72) $R = x_{max} - x_{min}$

73) $IQR = Q_3 - Q_1$

74) $\sigma_x = \sqrt{\frac{(x_1 - \bar{x})^2 + (x_2 - \bar{x})^2 + \cdots + (x_N - \bar{x})^2}{N - 1}}$ (there are a couple of variations of this formula)

75) $P = n!$

76) $P = \frac{n!}{r_1! r_2! \dots r_k!}$

77) $^n P_r = \frac{n!}{(n-r)!}$

78) $\binom{n}{r} = \frac{n!}{r!(n-r)!}$

79) $\Pr(A_i|B) = \frac{\Pr(B|A_i)\Pr(A_i)}{\Pr(B|A_1)\Pr(A_1) + \Pr(B|A_2)\Pr(A_2) + \cdots + \Pr(B|A_n)\Pr(A_n)}$

80) % error $= \frac{|obs. - acc.|}{acc.} 100\%$, % diff. $= \frac{|A-B|}{\frac{A+B}{2}} 100\%$

(an alternative is % diff. $= \frac{|A-B|}{\text{smaller of } A \text{ and } B} 100\%$)

81) $\chi^2 = \sum \frac{(O_i - E_i)^2}{E_i}$

82) $f(x) = \frac{e^{-(x-\mu)^2/2\sigma^2}}{\sqrt{2\pi\sigma^2}}$

83) $\sin\theta = \frac{\text{opp.}}{\text{hyp.}}$

84) $\cos\theta = \frac{\text{adj.}}{\text{hyp.}}$

85) $\tan\theta = \frac{\text{opp.}}{\text{adj.}}$

86) $\csc\theta = \frac{\text{hyp.}}{\text{opp.}}$

87) $\sec\theta = \frac{\text{hyp.}}{\text{adj.}}$

88) $\cot\theta = \frac{\text{adj.}}{\text{opp.}}$

89) $\tan\theta = \frac{\sin\theta}{\cos\theta}$

90) $\sin^2\theta + \cos^2\theta = 1, \sec^2\theta = 1 + \tan^2\theta$,

$\csc^2\theta = 1 + \cot^2\theta$

91) $\sin(\theta \pm \varphi) = \sin\theta\cos\varphi \pm \cos\theta\sin\varphi$

92) $\cos(\theta \pm \varphi) = \cos\theta\cos\varphi \mp \sin\theta\sin\varphi$

93) $\tan(\theta \pm \varphi) = \frac{\tan\theta \pm \tan\varphi}{1 \mp \tan\theta\tan\varphi}$

94) $\sin(2\theta) = 2\sin\theta\cos\varphi$

95) $\cos(2\theta) = \cos^2\theta - \sin^2\theta = 1 - 2\sin^2\theta = 2\cos^2\theta - 1$

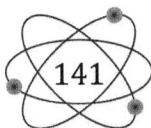

96) $\tan(2\theta) = \frac{2\tan\theta}{1-\tan^2\theta}$

97) $\sin\left(\frac{\theta}{2}\right) = \pm\sqrt{\frac{1-\cos\theta}{2}}$

98) $\cos\left(\frac{\theta}{2}\right) = \pm\sqrt{\frac{1+\cos\theta}{2}}$

99) $c^2 = a^2 + b^2 - 2ab\cos C$

100) $\frac{a}{\sin A} = \frac{b}{\sin B} = \frac{c}{\sin C} = 2R$

101) $\frac{a-b}{a+b} = \frac{\tan\left(\frac{A-B}{2}\right)}{\tan\left(\frac{A+B}{2}\right)}$

102) $A = \sqrt{s(s-a)(s-b)(s-c)}$ where

$s = \frac{a+b+c}{2}$

103) $\vec{A} = A_x\hat{i} + A_y\hat{j} + A_z\hat{k}$

104) $\|\vec{A}\| = \sqrt{A_x^2 + A_y^2 + A_z^2}$

105) $A_x = A\cos\theta, A_y = A\sin\theta$

106) $\theta = \tan^{-1}\left(\frac{A_y}{A_x}\right)$

107) $\vec{r} = x\hat{i} + y\hat{j} + z\hat{k}$

108) $\vec{v} = \frac{d\vec{r}}{dt}$

109) $\vec{a} = \frac{d\vec{v}}{dt} = \frac{d^2\vec{r}}{dt^2}$

110) $\Delta x = v_{x0}t + \frac{1}{2}a_x t^2, v_x = v_{x0} + a_x t,$

$v_x^2 = v_{x0}^2 + 2a_x\Delta x$ (notation varies, but check your

t's, squares, and initial values carefully)

111) $f = \frac{1}{T}$

112) $\omega = 2\pi f = \frac{2\pi}{T}$

113) $v = \lambda f = \frac{\lambda}{T}$ (bonus for $v = \frac{\omega}{k}$)

114) $\vec{A} \cdot \vec{B} = A_x B_x + A_y B_y + A_z B_z = AB\cos\theta$

115) $\vec{A} \times \vec{B} = \begin{vmatrix} \hat{i} & \hat{j} & \hat{k} \\ A_x & A_y & A_z \\ B_x & B_y & B_z \end{vmatrix}$

116) $\|\vec{A} \times \vec{B}\| = AB\sin\theta$

117) $W = \int_i^f \vec{F} \cdot d\vec{s}$ which simplifies to $W = \vec{F} \cdot \vec{s} =$

$Fs\cos\theta$ for a constant force

118) $\vec{\tau} = \vec{r} \times \vec{F}$

119) $\vec{v} = \vec{\omega} \times \vec{r}$

120) $\vec{A} \times (\vec{B} \times \vec{C}) = \vec{B}(\vec{A} \cdot \vec{C}) - \vec{C}(\vec{A} \cdot \vec{B})$

121) $\vec{A} \cdot (\vec{B} \times \vec{C}) = \vec{B} \cdot (\vec{C} \times \vec{A}) = \vec{C} \cdot (\vec{A} \times \vec{B}) =$

$\begin{vmatrix} A_x & A_y & A_z \\ B_y & B_y & B_z \\ C_x & C_y & C_z \end{vmatrix}$

122) $\vec{A} \times (\vec{B} \times \vec{C}) + \vec{B} \times (\vec{C} \times \vec{A}) + \vec{C} \times (\vec{A} \times \vec{B}) = 0$

123) $\cos^2\alpha + \cos^2\beta + \cos^2\gamma = 1$

124) $x' = x\cos\theta - y\sin\theta, y' = x\sin\theta + y\cos\theta$

125) $\tan\theta = \frac{m_1 - m_2}{1 + m_1 m_2}$

126) $\cos\theta = \frac{\vec{A} \cdot \vec{B}}{\|\vec{A}\|\|\vec{B}\|}$

127) $\log_b(xy) = \log_b x + \log_b y$

128) $\log_b(x^{-1}) = -\log_b x$

129) $\log_b(x/y) = \log_b x - \log_b y$

130) $\log_b(x^y) = y\log_b x$

131) $\ln(e^x) = x$

132) $\log_b x = \frac{\log_a x}{\log_a b}$

133) $\cosh x = \frac{e^x + e^{-x}}{2}$

134) $\sinh x = \frac{e^x - e^{-x}}{2}$

135) $\tanh x = \frac{\sinh x}{\cosh x} = \frac{e^x - e^{-x}}{e^x + e^{-x}}$

136) $f(x) \approx f(x_0) + \frac{x - x_0}{x_1 - x_0}[f(x_1) - f(x_0)]$

137) $f'(a) = \lim_{h \to 0} \frac{f(a+h) - f(a)}{h}$

138) $e = \lim_{n \to \infty} \left(1 + \frac{1}{n}\right)^n$

139) $\frac{d}{dx}x^k = kx^{k-1}$

140) $\frac{d}{dx}\sin x = \cos x, \frac{d}{dx}\cos x = -\sin x,$

$\frac{d}{dx}\tan x = \sec^2 x$

141) $\frac{d}{dx}\csc x = -\csc x\cot x, \frac{d}{dx}\sec x =$

$\sec x\tan x, \frac{d}{dx}\cot x = -\csc^2 x$

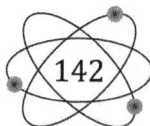

142) $\frac{d}{dx}\sin^{-1}x = \frac{1}{\sqrt{1-x^2}}, \frac{d}{dx}\cos^{-1}x = -\frac{1}{\sqrt{1-x^2}},$

$\frac{d}{dx}\tan^{-1}x = \frac{1}{1+x^2}$

143) $\frac{d}{dx}e^{kx} = ke^{kx}$

144) $\frac{d}{dx}\ln x = \frac{1}{x}$

145) $\frac{d}{dx}k^x = k^x \ln k$

146) $\frac{d}{dx}\sinh x = \cosh x, \frac{d}{dx}\cosh x = \sinh x,$

$\frac{d}{dx}\tanh x = \text{sech}^2 x$

147) $\frac{d}{dx}f(x)g(x) = f(x)\frac{dg(x)}{dx} + g(x)\frac{df(x)}{dx}$

148) $\frac{d}{dx}\frac{f(x)}{g(x)} = \frac{g(x)f'(x) - f(x)g'(x)}{g(x)^2}$

149) $\frac{d}{dx}f(g(x)) = \frac{df}{dg}\frac{dg}{dx}$

150) $\frac{d^n}{dx^n}f(x)g(x) = \sum_{k=0}^{n}\binom{n}{k}f^k(x)g^{n-k}(x)$

151) $A = \int_{x=a}^{b}f(x)\,dx$

152) $\int x^k\,dx = \frac{x^{k+1}}{k+1} + c$ if $k \neq -1$

153) $\int \frac{dx}{x} = \ln|x| + c$

154) $\int \sin x\,dx = -\cos x + c,$

$\int \cos x\,dx = \sin x + c$

155) $\int \tan x\,dx = \ln|\sec x| + c$ (same as

$-\ln|\cos x|), \int \cot x\,dx = \ln|\sin x| + c$

156) $\int \csc x\,dx = \ln\left|\tan\left(\frac{x}{2}\right)\right| + c$ (same as

$\ln|\csc x - \cot x|), \int \sec x\,dx = \ln|\sec x + \tan x| + c$

157) $\int e^{kx}\,dx = \frac{e^{kx}}{k} + c$ if $k \neq 0$

158) $\int \ln x\,dx = x\ln x - x + c$

159) $\int k^x\,dx = \frac{k^x}{\ln k}$ if $k > 0$ and $k \neq 1$

160) $\int \sinh x\,dx = \cosh x + c, \int \cosh x\,dx =$

$\sinh x + c$

161) $\int u\,dv = uv - \int v\,du$

162) $s = \int_i^f \sqrt{1 + \left(\frac{dy}{dx}\right)^2}\,dx$

163) $\vec{\mathbf{r}}_{cm} = \frac{\sum m_i \vec{\mathbf{r}}_i}{\sum m_i}$

164) $\vec{\mathbf{r}}_{cm} = \frac{1}{m}\int \rho\vec{\mathbf{r}}\,dV$ where $m = \int \rho\,dV$

165) $I = \sum m_i r_i^2$ (bonus for the formulas for the inertia tensor)

166) $I = \int \rho r_\perp^2\,dV$ (bonus for the formulas for the inertia tensor)

167) $(1+x)^k = 1 + \frac{kx}{1!} + \frac{k(k-1)x^2}{2!} +$

$\frac{k(k-1)(k-2)x^3}{3!} + \cdots + \frac{k(k-1)(k-2)\cdots(k-n+1)x^n}{n!} + \cdots$

where $|x| < 1$

168) $\Pr(a \leq X \leq b) = \int_a^b f(x)\,dx$

169) $\nabla f = \hat{\mathbf{i}}\frac{\partial f}{\partial x} + \hat{\mathbf{j}}\frac{\partial f}{\partial y} + \hat{\mathbf{k}}\frac{\partial f}{\partial z}$

170) $\nabla \cdot \vec{\mathbf{A}} = \frac{\partial A_x}{\partial x} + \frac{\partial A_y}{\partial y} + \frac{\partial A_z}{\partial z}$

171) $\nabla \times \vec{\mathbf{A}} = \begin{vmatrix} \hat{\mathbf{i}} & \hat{\mathbf{j}} & \hat{\mathbf{k}} \\ \frac{\partial}{\partial x} & \frac{\partial}{\partial y} & \frac{\partial}{\partial z} \\ A_x & A_y & A_z \end{vmatrix}$

172) $\nabla^2 f = \frac{\partial^2 f}{\partial x^2} + \frac{\partial^2 f}{\partial y^2} + \frac{\partial^2 f}{\partial z^2}$

173) $\Box f = \frac{1}{c^2}\frac{\partial^2 f}{\partial t^2} - \frac{\partial^2 f}{\partial x^2} - \frac{\partial^2 f}{\partial y^2} - \frac{\partial^2 f}{\partial z^2}$ (equivalent to $\partial^\mu \partial_\mu$ or $g^{\mu\nu}\partial_\mu \partial_\nu$)

174) $z = x + iy$

175) $z^* = x - iy$

176) $|z|^2 = z^*z = x^2 + y^2$

177) $z = |z|e^{i\theta}$

178) $e^{i\theta} = \cos\theta + i\sin\theta$

179) $(\cos\theta + i\sin\theta)^n = \cos(n\theta) + i\sin(n\theta)$

180) $z^{-1} = \frac{x}{x^2+y^2} - \frac{iy}{x^2+y^2}$ (equivalent to $\frac{e^{-i\theta}}{|z|}$)

181) $q = a + bi + cj + dk, i^2 = j^2 = k^2 = -1,$

$ij = -ji = k, jk = -kj = 1, ki = -ik = j$

182) $\int_C f(z)\,dz = 0$

183) $\det(A - \lambda I) = 0$

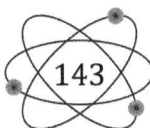

184) $A|a\rangle = \lambda|a\rangle$

185) $A^{-1} = \frac{1}{ad-bc}\begin{pmatrix} d & -b \\ -c & a \end{pmatrix}$

186) $[A,B] = AB - BA$, $\{A,B\} = AB + BA$

187) $[A,[B,C]] + [B,[C,A]] + [C,[A,B]] = 0$

188) $\kappa = \frac{y''}{(1+y'^2)^{3/2}}$

189) $\frac{dx}{dt} = \alpha x - \beta xy$, $\frac{dy}{dt} = \delta xy - \gamma y$

190) $\iiint_V \nabla \cdot \vec{F}\, dV = \oiint_S \vec{F} \cdot \hat{n}\, dA$

191) $\oint_C \vec{F} \cdot d\vec{s} = \iint_S (\nabla \times \vec{F}) \cdot d\vec{A}$ (equivalent to $\int_M d\omega = \int_{\partial M} \omega$)

192) $\oint_C P\,dx + Q\,dy = \iint_S \left(\frac{\partial Q}{\partial x} - \frac{\partial P}{\partial y} \right) dx\,dy$

193) $\nabla^2 f = 0$ (equivalent to $\frac{\partial^2 f}{\partial x^2} + \frac{\partial^2 f}{\partial y^2} + \frac{\partial^2 f}{\partial z^2} = 0$)

194) $\frac{\sqrt{2\pi n}}{n!} \left(\frac{n}{e} \right)^n \to 1$ as n grows to infinity

195) $x = r \cos\theta$, $y = r \sin\theta$

196) $r = \sqrt{x^2 + y^2}$, $\theta = \tan^{-1}\left(\frac{y}{x} \right)$

197) $x = r \cos\varphi \sin\theta$, $y = r \sin\varphi \sin\theta$, $z = r \cos\theta$ (note that many texts swap the roles of θ and φ)

198) $\hat{r} = \hat{i} \cos\varphi \sin\theta + \hat{j} \sin\varphi \sin\theta + \hat{k} \cos\theta$,

$\hat{\theta} = \hat{i} \cos\varphi \cos\theta + \hat{j} \sin\varphi \cos\theta - \hat{k} \sin\theta$,

$\hat{\varphi} = -\hat{i} \sin\varphi + \hat{j} \cos\varphi$ (note that many texts swap the roles of θ and φ)

199) $dV = r^2 \sin\theta\, dr\,d\theta\,d\varphi$

200) $\nabla f = \hat{r} \frac{\partial f}{\partial r} + \hat{\theta} \frac{1}{r} \frac{\partial f}{\partial \theta} + \hat{\varphi} \frac{1}{r \sin\theta} \frac{\partial f}{\partial \varphi}$

201) $\nabla \cdot \vec{A} = \frac{1}{r^2} \frac{\partial}{\partial r}(A_r r^2) + \frac{1}{r \sin\theta} \frac{\partial}{\partial \theta}(A_\theta \sin\theta) + \frac{1}{r \sin\theta} \frac{\partial A_\varphi}{\partial \varphi}$

202) $\nabla \times \vec{A} = \begin{vmatrix} \frac{\hat{r}}{r^2 \sin\theta} & \frac{\hat{\theta}}{r \sin\theta} & \frac{\hat{\varphi}}{r} \\ \frac{\partial}{\partial r} & \frac{\partial}{\partial \theta} & \frac{\partial}{\partial \varphi} \\ A_r & r A_\theta & r \sin\theta\, A_\varphi \end{vmatrix}$

203) $\gamma = \frac{1}{\sqrt{1 - \frac{v^2}{c^2}}}$

204) $t' = \gamma\left(t - \frac{vx}{c^2} \right)$, $x' = \gamma(x - vt)$, $y' = y$, $z' = z$

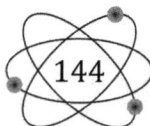

5 Pattern Challenge

1) 36, 43 (add 7)

2) 125, 71 (subtract 18)

3) I, K (skip one letter)

4) v, u (reverse order)

5) d, n (every other consonant)

6) V, M (reverse, skip two)

7) 17, 23 (add 1, add 2, add 3, etc.)

8) 140, 116 (minus 32, minus 30, minus 28, etc.)

9) p, q (letters with loops)

10) 29, 47 (add the previous two numbers)

11) 33, 12 (subtract the previous two numbers)

12) 99, 182 (add the previous three numbers)

13) 1875, 9375 (multiply by 5)

14) 324, 36 (divide by 3)

15) 1440, 10,080 (times 2, times 3, times 4, etc.)

16) 48, 1 (divide by 12, by 10, by 8, etc.)

17) 6, 24 (factorials)

18) 121, 169 (square odd numbers)

19) 125, 8 (17 cubed, 14 cubed, 11 cubed, etc.)

20) 7776, 117,649 (2^1, 3^2, 4^3, 5^4, etc.)

21) V, C (single-digit Roman numerals)

22) 1, 1111 (consecutive odd numbers in binary)

23) ty, ee (last two letters of 3, 6, 9, 12, 15, etc.)

24) 29, 31 (prime numbers)

25) 14, 21 (add 7, subtract 3, add 7, subtract 3, etc.)

26) 159, 477 (times 3, add 6, times 3, add 6, etc.)

27) 475, 955 (multiply by 2 and add 5)

28) 128, 26 (two series merged: 6, 10, 14, 18, 22, 26 adds 4, while 8, 16, 32, 64, 128 multiplies by 2)

29) 18, 25 (two series merged: 4, 5, 7, 10, 14, 19, 25 adds 1, 2, 3, etc., while 3, 6, 9, 12, 15, 18 adds 3)

30) 87, 20 (two series merged: 62, 55, 48, 41, 34, 27, 20 subtracts 7, while 47, 55, 63, 71, 79, 87 adds 8)

31) 3402, 3465 (alternately add and multiply the previous two numbers)

32) 2134, 2143 (rearrange the digits, put the 4-digit numbers in order)

33) 6P, 7N (add 1 to the number, skip one letter in reverse)

34) cPem, asBn (first letter skips one letter in reverse, M, k, i, g, E, c, a; second letter skips two letters, a, D, g, j, m, P, s; third letter skips two letters in reverse, t, q, N, k, h, e, B; last letter progresses naturally, h, i, j, K, l, m, n; the position of the uppercase letter advances forward each time)

35) 3:00, 4:45 (add 1 hour and 45 minutes)

36) lxiv, lxxxi (perfect squares in lowercase Roman numerals: 4, 9, 16, 25, 36, 49, 64, 81)

37) 344, 513 (add one to perfect cubes)

38) 28, 30 (subtract one from prime numbers)

39) 46,656, 823,543 (1^1, 2^2, 3^3, 4^4, etc.)

40) LIII, LXI (every other prime number in Roman numerals: 3, 7, 13, 19, 29, 37, 43, 53, 61)

41) 38, 3F (add 7 in the hexadecimal number system)

Note: If you think it should be 39 instead of 38, your mistake is that you're trying to add 8 to ordinary decimal numbers, when you should really be adding 7 to hexadecimal numbers. In the pattern, 31, 39 is the first time that two consecutive values don't contain a letter, whereas the part of the pattern 7, 15, 23, 31 always has a letter as one of the

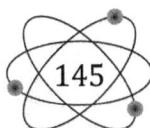

digits in between, which is why if you were trying to deduce the number following 31 without knowing about hexadecimal numbers, you could easily make the mistake of writing 39 instead of 38. If necessary, go online and convert 7, 14, 21, 28, 35, 42, 49, and 56 from decimal to hexadecimal to see that the correct answers are indeed 38 and 3F.

42) $10\frac{1}{8}$, $30\frac{3}{8}$ (multiply by 3)

43) $\frac{1}{2}$, 0.625 (add $\frac{1}{8}$, alternating between fraction, decimal, and percent)

44) $\frac{7}{4}$, $\frac{17}{6}$ (add the previous two fractions)

45) 0.015625, 0.0078125 (divide by 2)

46) 1296, 1944 (multiply by 1.5)

47) $\frac{35}{17}$, $\frac{39}{19}$ (denominators are prime numbers; to get the numerator, multiply the denominator by two and add one)

48) $\frac{15}{2}$, 7 (subtract $\frac{1}{2}$, multiply by 3, subtract $\frac{1}{2}$, multiply by 3, etc.)

49) $\frac{7}{12}$, $\frac{5}{12}$ (subtract $\frac{1}{12}$; the pattern is $\frac{10}{12}$, $\frac{9}{12}$, $\frac{8}{12}$, $\frac{7}{12}$, etc., except that each fraction is reduced when possible)

50) 528, 840 (square a prime number and subtract one)

51) 80,640, 40,320 (multiply by $\frac{8}{9}$, multiply by $\frac{7}{8}$, multiply by $\frac{6}{7}$, etc.)

52) Top right: add 2 to the top left. Bottom left: multiply the top squares. Bottom right: subtract the top right from the bottom left.

18	20
360	340

53) Top right: greatest common factor of the left squares. Bottom right: divide the bottom left by the top right.

180	180
360	2

54) Bottom right: find the positive difference between the top right and bottom left, and multiply this by the top left: $(30-10)\times 25 = 20\times 25 = 500$.

25	30
10	500

55) This puzzle rotates: the seed square is in the top left of the first array, the bottom left of the second array, the bottom right of the third array, etc. Multiply the seed number by the number counterclockwise from it, and add the number counterclockwise from that: $15\times 20 + 100 = 300 + 100 = 400$.

15	400
20	100

56) Top left: subtract the bottom left from the top right, then add the bottom right, and square the answer: $(20-10+15)^2 = 25^2 = 625$.

625	20
10	15

57) Advance the gray block 3 squares clockwise. The black blocks move diagonally down to the left, down to the right, and up to the right.

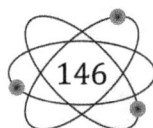

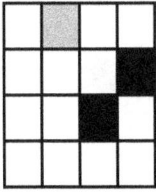

58) Advance the black block 2 squares clockwise. One white block advances 2 squares counterclockwise, while the other white block advances 3 squares clockwise.

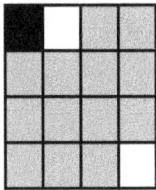

59) The black block advanced down, right, down, right, etc. A new gray block appears to take the place of the previous position of the black block.

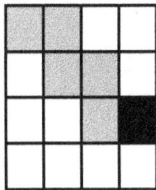

60) Rotate 90° counterclockwise. Parts of the black plus sign will fall off the grid during the rotation.

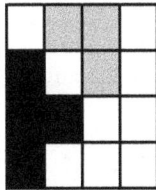

61) Advance the black block 1 square counterclockwise. One gray square advances 3 squares counterclockwise, while the other gray block advances 2 squares clockwise.

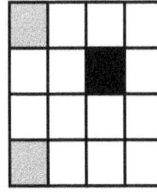

62) Rotate 90° counterclockwise.

63) Rhombus, star, rhombus, rhombus, star, rhombus, rhombus, star. Alternate gray and white.

64) One arrow rotates 45° clockwise. The other arrow rotates 135° clockwise. Occasionally, the two arrows happen to coincide.

65) Rotate 90° clockwise.

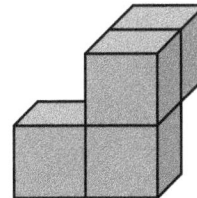

66) Rotate 90° about the vertical axis.

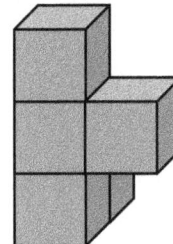

6 Test Your Math Skills

1) 7 (add 9, 7, and 5, subtract 6 and 8)

2) –56, 36, –7, 8 (two minus signs make a plus sign)

3) 144, 81, –64, 5, 1 and 128 (an even power of a negative value is positive, unlike an odd power)

4) 7 (exponent first, then multiply and divide, and then add and subtract)

5) 8 (add first, then divide, and then multiply)

6) 2 (subtract first, then add, and then divide)

7) ±7, ±14, ±1000

8) –4, ±2

9) 2×3×3×5 (or $2×3^2×5$)

10) 36

11) 80

12) $6\sqrt{2}$ (since 72 = 36×2)

13) 3285

14) 2804 (check: 2804+437 = 3241)

15) 6083

16) 1036

17) 79 (check: 79×24 = 1896)

18) 36 R 17 (check: 36×42 = 1512 and 1529 – 1512 = 17)

19) $\frac{4}{5}$

20) $\frac{23}{5}$

21) $\frac{11}{12}$

22) $\frac{11}{24}$

23) $\frac{1}{12}$

24) $\frac{2}{3}$

25) $\frac{1}{4}, \frac{1}{9}, \frac{9}{16}, \frac{125}{8}$

26) 16

27) $\pm\frac{1}{8}$ (same as ±0.125)

28) $\frac{\sqrt{3}}{3}$

29) 100,000, 0.01, 0.0001

30) 673.25

31) 7.022

32) 7.93

33) 22.68

34) 2.24

35) 36%, $\frac{9}{25}$

36) 0.625, 62.5%

37) 0.6, $\frac{3}{5}$

38) $0.\overline{75}$

39) $\frac{16}{33}$

40) 157% = 1.57, $\frac{8}{5}$ = 1.6, 1.65, $1\frac{2}{3}$ = $1.\overline{6}$ (where $1.\overline{6}$ means 1.66666666... with the 6 repeating forever)

41) 75 (check: 75 girls + 60 boys = 135 students and $\frac{75÷15}{60÷15} = \frac{5}{4}$ = 5:4)

42) 20 m/s (multiply by 1000 for kilo, and divide by 3600 to convert from per hour to per second)

43) 85 km/hr (use $\bar{v} = \frac{d}{t}$)

44) 140 km (check: $v = \frac{d}{t} = \frac{140}{4}$ = 35 km/hr)

45) 36 min. (check: 36 min. = 0.6 hours and $v = \frac{d}{t} = \frac{24}{0.6}$ = 40 km/hr)

46) $x = 8$ (check: 3×8–6 = 24–6 = 18)

47) $x = 2$ (check: 8–2×2 = 8–4 = 4 and 5×2–6 = 10–6 = 4)

48) x^{10} (note that $x^9 x^3 = x^{12}$ and $x^7 x^{-5} = x^2$)

49) $16x^{12}$

50) $12x^5 + 20x^4 - 8x^3$

51) $4x^3(2x^2 - 3)$ (you can check by distributing)

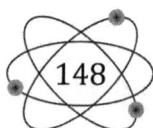

148

52) $8x^2 + 2x - 15$ (since $12x - 10x = 2x$)

53) $9 - x$ (note that $\sqrt{x}\sqrt{x} = x$)

54) $\frac{\sqrt{2x}}{x}$

55) $x = 6$ (check: $\frac{9}{6} = \frac{3}{2}$)

56) $x = 4$ (check: $\frac{1}{3} - \frac{1}{4} = \frac{4}{12} - \frac{3}{12} = \frac{1}{12}$)

57) no solution (since x cancels and $-3 \neq 5$)

58) all real numbers (since $3x + 4 = 3x + 4$)

59) $x = 0$ or 4 (check: $0^2 - 4 \times 0 = 0$ and $4^2 - 4 \times 4 = 16 - 16 = 0$)

60) $x = \pm 3$ (check: $2 \times 3^2 - 18 = 18 - 18 = 0$; same is true for -3 since the minus sign gets squared)

61) $x = 3$ (check: $3^2 - 6 \times 3 + 9 = 9 - 18 + 9 = 18 - 18 = 0$)

62) $x = -6$ or $\frac{3}{2}$ (check: $2 \times (-6)^2 + 9(-6) = 2 \times 36 - 54 = 72 - 54 = 18$ and $2\left(\frac{3}{2}\right)^2 + 9\left(\frac{3}{2}\right) = 2\left(\frac{9}{4}\right) + \frac{27}{2} = \frac{9}{2} + \frac{27}{2} = \frac{36}{2} = 18$)

63) $x = 8, y = 5$ (check: $3 \times 8 - 4 \times 5 = 24 - 20 = 4$ and $5 \times 8 + 2 \times 5 = 40 + 10 = 50$)

64) $x > -5$ or $-5 < x$ (check: plug -4.9 in for x to see that $2 - (-4.9) = 2 + 4.9 = 6.9 < 7$, and plug -5.1 in to see that $2 - (-5.1) = 2 + 5.1 = 7.1$)

65) $4 > x$ or $x < 4$ (check: plug 3.9 for x to see that $9 + 3.9 = 12.9 > 1 + 3(3.9) = 1 + 11.7 = 12.7$, that is $12.9 > 12.7$, and plug 4.1 in to get $9 + 4.1 = 13.1$ compared to $1 + 3(4.1) = 1 + 12.3 = 13.3$)

66) $36°$

67) $152°$

68) $42°$

69) $135°$

70) $143°$

71) $66°$

72) 25 units (since $225 + 400 = 625$)

73) 5 units (since $25 + 144 = 169$)

74) 5 units

75) 4 units

76) 30 units, 54 square units

77) 16 square units

78) 45 square units

79) $6\pi \approx 18.8$ units, $9\pi \approx 28.3$ square units

80) 17 units (apply Thales's theorem)

81) $2\pi \approx 6.28$ units (first convert $60°$ to $\frac{\pi}{3}$ radians), $30°$ (apply the inscribed angle theorem)

82) $y = 0.5x + 1.5$ (check: $0.5 \times 9 + 1.5 = 4.5 + 1.5 = 6$ and $0.5 \times (-7) + 1.5 = -3.5 + 1.5 = -2$)

83) 50,400 (use $\frac{10!}{3!3!2!}$)

84) 1:8 (there are $6^3 = 216$ possible outcomes, with 3 ways to get two 3's and one 4, 3 ways to get two 4's and one 2, 3 ways to get two 2's and one 6, 6 ways to get 2/3/5, 6 ways to get 1/3/6, and 6 ways to get 1/4/5; add these up to get 27 ways out of 216)

85) 4:9 (there are $\frac{9!}{2!7!} = 36$ ways to choose 2 balls out of 9; there are $\frac{5!}{2!3!} = 10$ ways to choose 2 blue balls out of 5; there are $\frac{4!}{2!2!} = 6$ ways to choose 2 red balls out of 4; and thus $\frac{10+6}{36} = \frac{16}{36} = \frac{4}{9}$ ways for both balls to have the same color)

86) 6 (the five data values have a sum of 30)

87) 7 (the value in the middle when ordered)

88) 5 (max minus min)

89) 2 (since $\sqrt{\frac{(3-6)^2 + (5-6)^2 + (7-6)^2 + (7-6)^2 + (8-6)^2}{5-1}}$ $= \sqrt{\frac{9+1+1+1+4}{4}} = \sqrt{\frac{16}{4}} = \sqrt{4} = 2$)

90) $\frac{2\pi}{3}$ rad

91) $135°$

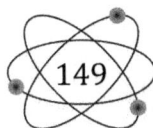

92) $\frac{\sqrt{3}}{2}, \frac{\sqrt{2}}{2}$ (same as $\frac{1}{\sqrt{2}}$), $\frac{\sqrt{3}}{3}$ (same as $\frac{1}{\sqrt{3}}$)

93) 2, $\sqrt{2}$ (note: $\frac{2}{\sqrt{2}} = \sqrt{2}$), $\sqrt{3}$ (note: $\frac{3}{\sqrt{3}} = \sqrt{3}$)

94) $\frac{1}{2}, -\frac{1}{2}, 0$

95) 150° (note that 30° doesn't lie in Quadrant II)

96) 180°

97) $\frac{2\sqrt{3}}{3}$, same as $\frac{2}{\sqrt{3}}$ (check with the law of sines:

$\frac{x}{\sin 45°} = \frac{2\sqrt{3}}{3} \div \frac{\sqrt{2}}{2} = \frac{2\sqrt{3}}{3} \times \frac{2}{\sqrt{2}} = \frac{2\sqrt{3}}{3} \times \sqrt{2} = \frac{2\sqrt{6}}{3}$ and

$\frac{\sqrt{2}}{\sin 60°} = \sqrt{2} \div \frac{\sqrt{3}}{2} = \sqrt{2} \times \frac{2}{\sqrt{3}} = \frac{2\sqrt{2}}{\sqrt{3}} = \frac{2\sqrt{6}}{3}$)

98) $\sqrt{3}$ (check with the law of cosines: $x^2 = 3 =$

$1 + 1 - 2\cos 120° = 2 - 2\left(-\frac{1}{2}\right) = 2 + 1 = 3$)

99) 60° or 300° (check: $6\cos 60° + 2 = 6\left(\frac{1}{2}\right) +$

$2 = 3 + 2 = 5$, same for 300°)

100) 30°, 60°, 210°, or 240° (check: $4\sin 30° \cos 60°$

$= 4\left(\frac{1}{2}\right)\left(\frac{\sqrt{3}}{2}\right) = \sqrt{3}$, same for the others; one way to

solve the problem is to use $\sin 2\theta = 2\sin\theta\cos\theta$)

101) 30°, 150°, 210°, or 330° (check: $9\sin^2 30° +$

$\cos^2 30° = 9\left(\frac{1}{2}\right)^2 + \left(\frac{\sqrt{3}}{2}\right)^2 = \frac{9}{4} + \frac{3}{4} = \frac{12}{4} = 3$; one

way to solve the problem is $\cos^2\theta = 1 - \sin^2\theta$)

102) 90° (check: $\sin^2 90° + \sin 90° - 2 = 1^2 + 1 -$

$2 = 0$; one way to solve the problem is to define

$x = \sin\theta$ and solve the quadratic equation to find

that $x = 1$ or $x = -2$, for which only $x = 1$ leads

to a real solution)

103) 0

104) 3

105) 1

106) –2

107) 2

108) 6

109) 78 (simply plug the value in)

110) $x \geq 2, f \leq 0$

111) $-\infty < x < \infty, f \geq 0$

112) 7

113) $\infty, -\infty$ (showing that $\lim_{x \to 0} \frac{1}{x}$ doesn't exist)

114) 3 (one way is to first divide the numerator and denominator both by x)

115) 4 (note that $x^2 - 4 = (x+2)(x-2)$; you don't need to worry about division by zero since this is a limit – that is, we're exploring what happens to the ratio as we get closer and closer to $x = 2$ without actually reaching this value)

116) 1 (apply l'Hôpital's rule, or make a graph)

117) $12x^2$

118) 1 (occasionally, good calculus students let themselves get fooled by such a simple problem; note that $x^1 = x$, so that $\frac{d}{dx}x = 1x^0 = x$)

119) $\frac{\sqrt{x}}{2x}$, same as $\frac{1}{2\sqrt{x}}$ or $2^{-1}x^{-1/2}$ (note that $\sqrt{x} = x^{1/2}$)

120) $24x(3x^2 - 5)^3$ (let $u = 3x^2 - 5$ and $f = u^4$, and apply the chain rule, $\frac{df}{dx} = \frac{df}{du}\frac{du}{dx}$)

121) $\frac{x^2+4x-6}{x^2+4x+4}$ (apply the quotient rule; note that $(x+2)^2 = x^2 + 4x + 4$)

122) $8\sin x \cos x$, same as $4\sin 2x$ (let $u = \sin x$ and $f = 4u^2$, and apply the chain rule, $\frac{df}{dx} = \frac{df}{du}\frac{du}{dx}$; recall that $\sin 2x = 2\sin x \cos x$)

123) $8x\cos(x^2)$ (let $u = x^2$ and $f = 4\sin u$, and apply the chain rule, $\frac{df}{dx} = \frac{df}{du}\frac{du}{dx}$)

124) $\sec x \tan x$, same as $\frac{\sin x}{\cos^2 x}$ (one way is to write $\sec x = \frac{1}{\cos x}$)

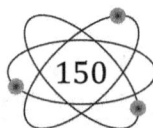

125) $-\frac{1}{\sqrt{1-x^2}}$ (let $u = \cos^{-1} x$ such that $\cos u =$ $\cos(\cos^{-1} x) = x$; take an implicit derivative to get $-\sin u \frac{du}{dx} = 1$, which becomes $\frac{du}{dx} = -\frac{1}{\sin u} = -\frac{1}{\sin(\cos^{-1} x)}$; recall that $\sin(\cos^{-1} x) = \sqrt{1-x^2}$)

126) $\sin x + x \cos x$ (apply the product rule)

127) $3e^{3x}$ (let $u = 3x$ and apply the chain rule)

128) $\frac{1}{x}$ (let $u = 3x$ and apply the chain rule; note that the 3 cancels out, in contrast to exercise 127, ultimately due to the property $\ln 3x = \ln 3 + \ln x$)

129) $2^x \ln 2$ (use $e^{\ln 2} = 2$ to write $2^x = \left(e^{\ln 2}\right)^x = e^{x \ln 2}$, then let $u = x \ln 2$ and $f = e^u$ to use the chain rule)

130) $x^x(\ln x + 1)$ (let $f = x^x$ and write $\ln f = \ln(x^x) = x \ln x$ to get $\frac{1}{f}\frac{df}{dx} = \ln x + 1$, such that $\frac{df}{dx} = f(\ln x + 1) = x^x(\ln x + 1)$)

131) $\frac{1}{x \ln 10}$ (use $\log_{10} x = \frac{\ln x}{\ln 10}$)

132) 5 (since $f' = \frac{df}{dx} = 2x - 3$)

133) $60x^2 - 24x + 6$ (since the first derivative is $20x^3 - 12x^2 + 6x - 2$)

134) relative minimum of –162 at $x = 3$, relative maximum of 162 at $x = -3$, absolute minimum of –162 at $x = 3$, absolute maximum of 1250 at $x = 5$ (the derivative is $5x^4 - 45x^2$; set this equal to zero to get $x = \pm 3$ or 0; evaluate the function at these values of x to determine the relative extrema; also evaluate the function at the endpoints)

135) $6x^4 + c$ (check: $\frac{d}{dx}6x^4 = 24x^3$)

136) $-\frac{1}{2x^2} + c$ (check: $-\frac{d}{dx}\frac{1}{2x^2} = \frac{1}{x^3}$; one way is to rewrite $\frac{1}{x^3}$ as x^{-3})

137) $\ln x + c$ (check: $\frac{d}{dx}\ln x = \frac{1}{x}$)

138) $\frac{2x\sqrt{x}}{3} + c$, same as $\frac{2x^{3/2}}{3}$ (check: $\frac{d}{dx}\frac{2}{3}x^{3/2} = \sqrt{x}$; one way is to rewrite $\sqrt{x}$ as $x^{1/2}$)

139) 1 (note that $\int \sin x \, dx = -\cos x + c$; evaluate the antiderivative over the endpoints to get $-\cos\frac{\pi}{2} + \cos 0 = -0 + 1 = 1$; a graph of sine over this interval should convince you that the area under the curve must be positive)

140) $\ln|\sec x| + c$, same as $-\ln|\cos x|$ (check: $\frac{d}{dx}\ln|\sec x| = \frac{1}{\sec x}\sec x \tan x = \tan x$; one way is to write $\tan x = \frac{\sin x}{\cos x}$, and make the substitution $u = \cos x$ and $du = -\sin x \, dx$)

141) $\ln|\sec x + \tan x| + c$ (check: $\frac{d}{dx}\ln|\sec x + \tan x| = \frac{1}{\sec x + \tan x}(\sec x \tan x + \sec^2 x) = \sec x$)

142) $-\frac{e^{-ax}}{a} + c$ (check: $-\frac{d}{dx}\frac{e^{-ax}}{a} = e^{-ax}$)

143) $x \ln x - x + c$ (check: $\frac{d}{dx}x \ln x - \frac{d}{dx}x = \ln x + \frac{x}{x} - 1 = \ln x$; one way is to integrate by parts with $u = \ln x$ and $dv = dx$, such that $uv - \int v \, du = x \ln x - \int x\frac{1}{x}dx = x \ln x - x$)

144) $\cosh x + c$ (check: $\frac{d}{dx}\cosh x = \sinh x$; note that there isn't a minus sign, in contrast to the analogous trig integral)

145) $-2\sqrt{1-x} + c$ (check: $-\frac{d}{dx}2\sqrt{1-x} = \frac{1}{\sqrt{1-x}}$; one way is to make the substitution $u = 1 - x$ and $du = -dx$)

146) $\sin^{-1} x + c$ (check: $\frac{d}{dx}\sin^{-1} x = \frac{1}{\sqrt{1-x^2}}$; one way is to make the substitution $x = \sin u$ and $dx = \cos u \, du$; note that $1 - \sin^2 u = \cos^2 u$)

147) $\frac{e^{x^2}}{2} + c$ (check: $\frac{d}{dx}\frac{e^{x^2}}{2} = xe^{x^2}$; one way is to make the substitution $u = x^2$ and $du = 2x \, dx$)

148) $e^x(x-1) + c$, same as $xe^x - e^x$ (check:

$\frac{d}{dx}xe^x - \frac{d}{dx}e^x = e^x + xe^x - e^x = xe^x$; one way is

to integrate by parts with $u = x$ and $dv = e^x dx$)

149) $\frac{x}{2} + \frac{\sin 2x}{4} + c$ (check: $\frac{d}{dx}\frac{x}{2} + \frac{d}{dx}\frac{\sin 2x}{4} = \frac{1}{2} +$

$\frac{1}{2}\cos 2x = \frac{1}{2} + \frac{1}{2}(\cos^2 x - \sin^2 x) = \frac{1}{2} + \frac{1}{2}\cos^2 x -$

$\frac{1}{2}\sin^2 x = \frac{1}{2}(1 - \sin^2 x) + \frac{1}{2}\cos^2 x = \frac{1}{2}\cos^2 x +$

$\frac{1}{2}\cos^2 x = \cos^2 x$; one way is to use the identity

$\cos^2 x = \frac{1+\cos 2x}{2}$)

150) $\sin x - \frac{\sin^3 x}{3} + c$ (check: $\frac{d}{dx}\sin x - \frac{d}{dx}\frac{\sin^3 x}{3} =$

$\cos x - \sin^2 x \cos x = \cos x - (1 -$

$\cos^2 x)\cos x = \cos x - \cos x + \cos^3 x = \cos^3 x$;

one way is to write $\cos^3 x = \cos^2 x \cos x$ and use

the identity $\cos^2 x = 1 - \sin^2 x$)

151) $\frac{\sec x \tan x}{2} + \frac{\ln|\sec x + \tan x|}{2} + c$ (check:

$\frac{d}{dx}\frac{\sec x \tan x}{2} + \frac{d}{dx}\frac{\ln|\sec x + \tan x|}{2} = \frac{\sec x \tan^2 x}{2} + \frac{\sec^3 x}{2} +$

$\frac{\sec x \tan x + \sec^2 x}{2(\sec x + \tan x)} = \frac{\sec x \tan^2 x}{2} + \frac{\sec^3 x}{2} + \frac{\sec x}{2} =$

$\frac{\sec x}{2}(\tan^2 x + 1) + \frac{\sec^3 x}{2} = \frac{\sec^3 x}{2} + \frac{\sec^3 x}{2} = \sec^3 x$;

one way is to integrate by parts with $u = \sec x$ and

$dv = \sec^2 x\, dx$, apply the identity $\tan^2 x + 1 =$

$\sec^2 x$, add $\int \sec^3 x\, dx$ to both sides of the

equation, and divide by two)

152) $\frac{\tan^3 x}{3} + \tan x + c$ (check: $\frac{d}{dx}\frac{\tan^3 x}{3} + \frac{d}{dx}\tan x =$

$\tan^2 x \sec^2 x + \sec^2 x = \sec^2 x(\tan^2 x + 1) =$

$\sec^2 x \sec^2 x = \sec^4 x$; one way is to rewrite $\sec^4 x$

as $\sec^2 x \sec^2 x$, apply $\sec^2 x = \tan^2 x + 1$ to one

of these to separate the integral into two terms, and

make the substitution $u = \tan x$ and $du =$

$\sec^2 x\, dx$)

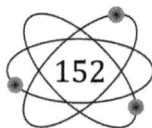

7 Axioms, Theorems, Laws, etc.

1) Transitive property: If $a = c$ and $b = c$, then $a = b$. (This extends beyond algebra. For example, there is a similar statement in set theory and in thermodynamics – the principle of thermometry.)

2) Reflexive property: $x = x$. (This is useful in geometry proofs and there is a similar relation for sets.)

3) Symmetric property: If $x = y$ then $y = x$.

4) Unique Factorization Theorem: Any integer greater than unity can be expressed as a unique product of prime numbers (that is, unique except for the order in which they are multiplied).

5) Prime Number Theorem: As x approaches infinity, the number of prime numbers less than or equal to x approaches the ratio $\frac{x}{\ln x}$ (Jacque Hadamard and Charles De La Vallée-Poussin, independently, 1896).

6) Chebyshev's Theorem: For a positive integer n, there exists a prime number between n and $2n$.

7) Bertrand's Postulate: For a positive integer $n > 3$, there exists a prime number between n and $2(n - 1)$.

8) Peano Axioms: (i) Natural numbers include zero. (ii) There is a successor for every natural number. (iii) Zero isn't the successor to any natural number. (iv) If m and n have the same successor, $m = n$. (v) If a set contains zero and also contains the successor of every number, the set contains the natural numbers. Note: These are sometimes presented in different forms (and sometimes with additional axioms).

9) Discriminant: $b^2 > 4ac$ gives two distinct real roots, $b^2 = 4ac$ gives a double root, and $b^2 < 4ac$ only has complex roots.

10) Remainder Theorem: A polynomial $f(x)$ divided by $x - h$ has a remainder equal to $f(h)$. Factor Theorem: $x - h$ is a factor of polynomial $f(x)$ if and only if $f(h) = 0$.

11) Fermat's Last Theorem: $x^n + y^n = z^n$ has no solutions where x, y, and z are all integers if n is an integer greater than two. Note: Be careful not to confuse this with Fermat's Little Theorem.

12) Four Squares Theorem: Any positive integer can be expressed as the sum of the squares of one, two, three, or four positive numbers. Lagrange's Theorem: Any positive integer can be expressed as $p = a^2 + b^2 + c^2 + d^2$, where a, b, c, and d are nonnegative numbers.

13) Four Color Theorem: A maximum of four colors are needed to color a map such that no two adjacent regions have the same color (proven using a computer by Kenneth Appel and Wolfgang Haken, 1976).

14) Handshaking Lemma: The sum of the degrees of all of the vertices of any graph is even.

15) Euclid's Axioms: (i) A straight line can be drawn to connect any two points. (ii) Any line segment can be extended infinitely in either direction. (iii) Any line segment can serve as the radius of a circle with one endpoint as its center. (iv) All right angles are congruent. (v) Two lines joined by a transversal intersect (if extended far enough) if the inner angles add up to less than two right angles.

16) Angle Sum Theorem: The angular measures of the three interior angles of any triangle add up to 180°.

17) Angle Sum Theorem for Polygons: The angular measures of the interior angles of any polygon add up to $180°(n-2)$, where n is the number of sides.

18) Triangle Inequality Theorem: The sum of the lengths of any two sides of a triangle is greater than the length of the remaining side.

19) Transversal/Parallel Lines: (i) Four pairs of corresponding angles are congruent. (ii) Two pairs of alternate interior angles are congruent. (iii) Two pairs of alternate exterior angles are congruent. (iv) Four pairs of vertical angles are also congruent.

20) Triangle Congruence: SSS (all three sides are congruent), SAS (two sides and the angle formed by those sides are congruent), ASA (two angles and the side that touches both angles are congruent), AAS (two angles and a side that doesn't touch both angles are congruent).

21) Triangle Similarity: AA (two angles are congruent), or all three sides come in the same proportions.

22) Perpendicular Bisector Equidistant Theorem: Any point that lies on the perpendicular bisector of a line segment is equidistant from the endpoints of the line segment.

23) Triangle Bisector Theorem: The line that bisects an angle of a triangle divides the opposite edge of the triangle into segments in proportion to the lengths of the other two sides.

24) Midsegment Theorem: The midsegment, which connects the midspoints of two sides of a triangle, is parallel to the remaining side of the triangle.

25) Inscribed Angle Theorem: An angle inscribed in a circle has an angular measure equal to one-half of the angular measure of the central angle that intercepts the same arc length.

26) Thales's Theorem: If a triangle is inscribed in a circle such that one side of the triangle is a diameter, the angle opposite to the diameter is a right angle.

27) Tangent-Chord Theorem: The angle between a chord and a line that is tangent to the circle at one of the chord's endpoints has an angular measure that is one-half of the central angle formed by the chord.

28) Ptolemy's Theorem: A quadrilateral with vertices in the order A, B, C, and D is cyclic (meaning that all four vertices lie on the same circle) if and only if $AB \cdot CD + AD \cdot BC = AC \cdot BD$.

29) Parallelogram Law: The sum of the squares of the lengths of the four sides of a parallelogram is equal to the sum of the squares of the lengths of the two diagonals. (This is the basis for vector addition.)

30) Morley's Theorem: The adjacent trisectors of the interior angles of any triangle intersect at points that form the vertices of an equilateral triangle.

31) Pappus's Centroid Theorem: If a plane region is rotated about a line within the plane (that doesn't cut the region) through one complete revolution, the volume of the solid of revolution is equal to the product

of the area of the region and the distance traveled by the centroid of the region.

32) Law of Averages: The incorrect notion that the outcome of an event has a higher probability of occurring if it has been under-observed thus far. (This is basically the Gambler's Fallacy.)

33) Law of Large Numbers: The mean value of the sample approaches the expected value as the number of identically distributed random variables increases (once the number is very large). It is instructive to think about #32 and #33 (then research the distinction, if necessary).

34) Central Limit Theorem: As the number of identically distributed random variables increases, the distribution of the mean value approaches the standard normal distribution.

35) Bernoulli's Theorem: For a sequence of independent trials with possible outcomes of success and failure, with a probability p for success, if there are m successful events in n trials, the relative frequency $\frac{m}{n}$ approaches p as the sample size approaches infinity (a special case of the Weak Law of Large Numbers).

36) Addition Law: The probability that event A or event B will occur equals the probability that event A will occur plus the probability that event B will occur minus the probability that both will occur.

37) Absorption Laws: $A \cap (A \cup B) = A$ and $A \cup (A \cap B)$, given any two subsets, A and B.

38) De Morgan's Laws: (i) The negation of a disjunction is equivalent to the conjunction of the negations. (ii) The negation of a conjunction is equivalent to the disjunctions of the negations.

39) Inclusion-Exclusion Principle: The cardinality of the union of n sets equals the cardinality of the sets minus the cardinalities of pairs of sets plus the cardinalities of triples of sets minus the cardinalities of quadruples of sets, etc. (continuing with alternating signs). It basically generalizes #36.

40) Osborne's Rule: The identities for hyperbolic functions, like $\sinh x$, are the same as the identities for the basic trig functions, like $\sin x$, except that a product of $\sinh$'s causes a negative sign.

41) Fundamental Theorem of Algebra: Any nonzero, single variable polynomial of degree n has exactly n roots (allowing for multiplicity), whether the coefficients are complex or real. (Another form of the rule replaces "nonzero" with "constant" and "exactly n roots" with "at least one root.")

42) Fundamental Theorem of Calculus: For any function $f(x)$ that is continuous on $[a, b]$ for which $\frac{dg}{dx} = f(x)$ for all x in the interval, $\int_{x=a}^{b} f(x)\, dx = g(b) - g(a)$. That is, a definite integral of a function $f(x)$ may be evaluated by finding the antiderivative, $g(x)$, evaluating it at the endpoints, and subtracting.

43) Mean Value Theorem: For any function $f(x)$ that is continuous on $[a, b]$ and differentiable on (a, b), there exists a number c, where $a < c < b$, for which $\frac{df}{dx}\Big|_{x=c} = \frac{f(b)-f(a)}{b-a}$.

44) Extreme Value Theorem: For any real function $f(x)$ that is continuous on $[a, b]$, there exist numbers c and d, where $a \leq c \leq b$ and $a \leq d \leq b$, for which $f(c) \leq f(x) \leq f(d)$ for all x in this interval.

45) Intermediate Value Theorem: For any real function $f(x)$ that is continuous on $[a, b]$, there exists a number g, where $f(a) \leq g \leq f(b)$, for which $f(c) = g$ for some value c where $a < c < b$.

46) L'Hôpital's Rule: If $f(x)$ and $g(x)$ both approach zero as x approaches a (such that the ratio $\frac{f}{g}$ appears to have an indeterminant form in this limit), the limit $\frac{f}{g}$ as x approaches a equals the limit of the ratio of $\frac{df}{dx}$ to $\frac{dg}{dx}$ evaluated at $x = a$.

47) Rolle's Theorem: For any function $f(x)$ that is continuous on $[a, b]$ and differentiable on (a, b) for which $f(a) = f(b)$, there exists a number c, where $a < c < b$, for which $\left.\frac{df}{dx}\right|_{x=c} = 0$ (which is thus a critical point, also called a stationary point). This is a special case of the Mean Value Theorem.

48) Derivative Test: The first derivative of a function equals zero at a critical point (also called a stationary point). If the second derivative evaluated at a critical point is negative, the critical point is a relative maximum; if the second derivative evaluated at a critical point is positive, the critical point is a relative minimum; if the second derivative evaluated at a critical point is zero, the critical point may be a point of inflection (but further analysis is required to determine this).

49) Abel's Test: If the infinite series $\sum a_n$ is convergent and $\{b_n\}$ is monotonically decreasing for all n, then the infinite series $\sum a_n b_n$ is also convergent.

50) Comparison Test: If $0 \leq a_i \leq b_i$ for all $i > j$ (where j is a positive integer), then if $\sum b_n$ converges $\sum a_n$ also converges, and if $\sum a_n$ diverges $\sum b_n$ also diverges.

51) Dirichlet's Test: If $\{a_n\}$ has bounded partial sums and if $\{b_n\}$ is decreasing and converges to zero, then the infinite series $\sum a_n b_n$ converges.

52) Ratio Test: If $\left|\frac{a_{n+1}}{a_n}\right|$ approaches r as n becomes infinite, then if $r < 1$ the infinite series converges absolutely, if $r > 1$ the series diverges, and if $r = 1$ further analysis is required.

53) Jordan Curve Theorem: A simple closed curve divides the plane into the two regions that are the interior and exterior of the curve. (Although the result seems obvious, the proof is nontrivial.)

54) Riemann Zeta Hypothesis: The Riemann zeta function has its only (nontrivial) zeroes for complex numbers for which the real part equals one-half.

8 Math History Challenge

1) Euclid, circa 300 BC
2) Pythagoras, 500's BC; however, the theorem may have been known long before his time
3) Blaise Pascal, 1600's; however, this triangle was known earlier in China and Persia
4) Leonhard Euler, 1700's
5) Muhammad ibn Mūsā al-Khwārizmī, circa 800, algebra
6) René Descartes (namely, "Cartesian"), 1600's
7) Archimedes, 200's BC
8) Fibonacci (a pseudonym of Leonardo Pisano), late 1100's and into the 1200's
9) Liu Hui, 200's BC
10) Isaac Newton, mid 1600 into the 1700's, flowing quantity, instantaneous rate of flow
11) Gottfried Wilhelm Leibniz, mid 1600 to early 1700's
12) Brahmagupta, 600's
13) Carl Friedrich Gauss, late 1700's to mid 1800
14) Thales of Miletus, late 600's to mid 500 BC
15) Pierre de Fermat, 1600's
16) Zu Chongzhi, 400's
17) (Ghiyāth al-Dīn Jamshīd Mas'ūd) al-Kāshi, late 1300's into the 1400's
18) Amalie Noether (known as Emmy), late 1800's into the 1900's
19) Eratosthenes, 200's BC
20) John Napier, mid 1500 into the 1600's
21) Jakob (aka Jacques) Bernoulli, late 1600's
22) Johann (aka Jean) Bernoulli, late 1600's
23) Apollonius, 200's BC and into the early 100's
24) Kurt Gödel, 1900's
25) Claudius Ptolemaeus (aka Ptolemy), 100's
26) Hipparchus, 100's BC
27) Srinivasa Ramanujan, late 1800's to early 1900's
28) Girolamo Cardano, 1500's
29) Rafael Bombelli, 1500's
30) Mikhail Leonidovich Gromov, 1943
31) Jia Xian (aka Chia Hsien), 1000's
32) Yang Hui (aka Qianguang), 1200's
33) Zhu Shijie (aka Hanqing, Songting, or Chu Shih-Chieh), mid 1200 into the 1300's
34) Omar Khayyam, mid 1000 into the 1100's
35) Maryam Mirzakhani, 1977
36) Georg Ferdinand Ludwig Philipp Cantor, mid 1800 into the 1900's
37) Grigori Perelman, 1966
38) Joseph Louis Lagrange, 1700's to early 1800's
39) John Forbes Nash, Jr., 1928
40) Marie-Sophie Germain, late 1700's into the 1800's
41) Andrew Wiles, 1953, his student, Richard Taylor
42) John Wallis, 1600's
43) John Willard Milnor, Pierre René Deligne, Jean-Pierre Serre, and John Griggs Thompson
44) Nicole (or Nicholas) Oresme, 1300's, Merton's theorem (or rule)
45) Alexander Grothendieck, 1928
46) Bonaventura Cavalieri, 1600's
47) Karen Keskulla Uhlenbeck, 1942
48) Caucher Birkar, Alessio Figalli, Peter Scholze, and Askhay Venkatesh

9 Math Abbreviation Challenge

1) No. = number
2) c. = circa
3) LCD = least common denominator, LCM = least common multiple (we'll accept "lowest")
4) GCF = greatest common factor, HCF = highest common factor, GCD = greatest common divisor
5) PEMDAS = parentheses exponents multiplication/division (in order) addition/subtraction (in order)
6) TI = Texas Instruments
7) CPA = certified public accountant
8) AR = accounts receivable, AP = accounts payable
9) CR = credit, DR = debit (an interesting abbrev.)
10) ROI = return on investment
11) IRA = individual retirement account
12) P&L = profit and loss
13) LTL = long-term liabilities
14) IRS = Internal Revenue Service
15) STEM = science, technology, engineering, math
16) M.S. = Master of Science, Ph.D. = Doctor of Philosophy
17) APR = annualized percentage rate
18) UPC = universal product code
19) RPI = retail price index
20) EMV = expected monetary value
21) ARM = adjustable rate mortgage
22) CD = certificate of deposit
23) NYSE = New York Stock Exchange
24) GNP = Gross National Product
25) EFT = electronic funds transfer
26) YTD = year-to-date
27) NASDAQ = National Association of Securities Dealers Automated Quotation
28) ASCII = American Standard Code for Information Interchange
29) IMO = International Math Olympiad
30) AMC = American Mathematics Competition
31) foil = first outside inside last
32) QE = quadratic equation
33) FLT = Fermat's Last Theorem
34) SSS = side side side, AA = angle angle

35) SAS = side angle side, ASA = angle side angle, AAS = angle angle side
36) CPCTC = corresponding parts of congruent triangles are congruent
37) PT = Pythagorean Theorem
38) PBET = Perpendicular Bisector Equidistant Theorem
39) i.e. = id est (that is), e.g. = exempli gratia (for example)
40) et al. = et alii (and others)
41) n.b. = nota bene (note well)
42) viz. = videlicet (namely)
43) vs. = versus, v.s. = vide supra (see above)
44) cf. = confer (compare)
45) wlog = without loss of generality
46) wrt = with respect to
47) iff = if and only if
48) STP = sufficient to prove
49) QED = quod erat demonstrandum, QEF = quod erat faciendum
50) SI = le Système International (d'unités), aka the International System (of units)
51) lb = libra (for pound)
52) oz = ounce
53) qt = quart, pt = pint, c = cup
54) tbsp = tablespoon, tsp = teaspoon
55) cc = cubic centimeters, L = liters
56) F = Fahrenheit, C = Celsius (or Centigrade), K = Kelvin
57) mph = miles per hour
58) rpm = revolutions (or rotations) per minute
59) c = centi (10^{-2}), d = deci (10^{-1}), da = deca (10^{1}), f = femto (10^{-15}), G = giga (10^{9}), h = hecto (10^{2}), k = kilo (10^{3}), m = milli (10^{-3}), μ = micro (10^{-6}), n = nano (10^{-9}), p = pico (10^{-12}), T = tera (10^{12})
60) AMS = American Mathematical Society
61) MAA = Mathematical Association of America
62) EMS = European Mathematical Society
63) JAMS = Journal of the American Mathematical Society
64) EJM = European Journal of Mathematics

65) Institute of Mathematics and its Applications

66) sohcahtoa = sine opposite/hypotenuse, cosine adjacent/hypotenuse, tangent opposite/adjacent

67) ASTC = all sine tangent cosine (there are multiple mnemonic acronyms to help remember this, such as "All students take calculus")

68) SHM = simple harmonic motion

69) exp = exponential function (e)

70) log = logarithm, ln = natural logarithm

71) DNE = does not exist

72) RSA = Rivest-Shamir-Adleman

73) rv = random variable

74) sd = standard deviation, se = standard error

75) rms = root-mean-square

76) IQR = interquartile range

77) p.d.f. = probability distribution function, p.g.f. = probability generating function

78) c.d.f. = cumulative distribution function

79) EDA = exploratory data analysis

80) i.i.d. = independent and identically distributed

81) erf = error function

82) pmcc = product moment correlation coefficient

83) FTOC = Fundamental Theorem of Calculus

84) MVT = Mean Value Theorem

85) sgn = signum function

86) DE = differential equation, ODE = ordinary differential equation, PDE = partial differential equation

87) BC = boundary condition

88) RKM = Runge Kutta method

89) TSP = traveling salesman problem

90) gf = generating function

91) real part, imaginary part

92) complex conjugate

93) adj. = adjoint or adjugate

94) tr = trace

95) dof = degrees of freedom

96) FFT = fast Fourier transform

97) SU = special unitary, SO = special orthogonal

98) SUSY = supersymmetry

99) TOE = theory of everything

100) GUT = grand unified theory

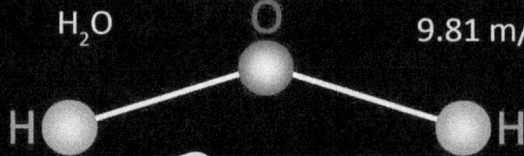

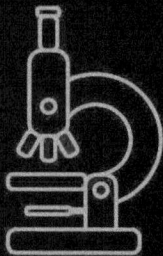

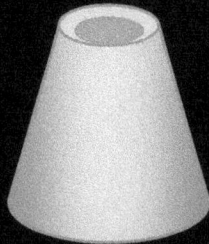

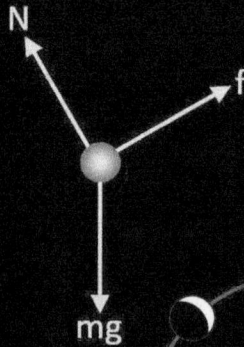

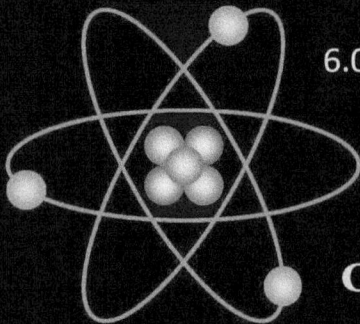

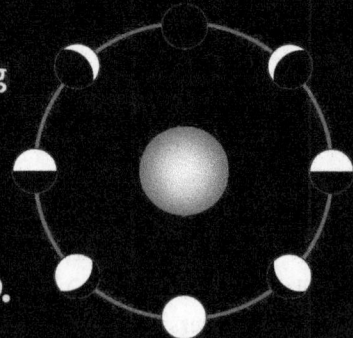

ABOUT THE AUTHOR

Dr. Chris McMullen has over 20 years of experience teaching university physics in California, Oklahoma, Pennsylvania, and Louisiana. Dr. McMullen is also an author of math and science workbooks. Whether in the classroom or as a writer, Dr. McMullen loves sharing knowledge and the art of motivating and engaging students.

The author earned his Ph.D. in phenomenological high-energy physics (particle physics) from Oklahoma State University in 2002. Originally from California, Chris McMullen earned his Master's degree from California State University, Northridge, where his thesis was in the field of electron spin resonance.

As a physics teacher, Dr. McMullen observed that many students lack fluency in fundamental math skills. In an effort to help students of all ages and levels master basic math skills, he published a series of math workbooks on arithmetic, fractions, long division, algebra, geometry, trigonometry, and calculus entitled *Improve Your Math Fluency*. Dr. McMullen has also published a variety of science books, including astronomy, chemistry, and physics workbooks.

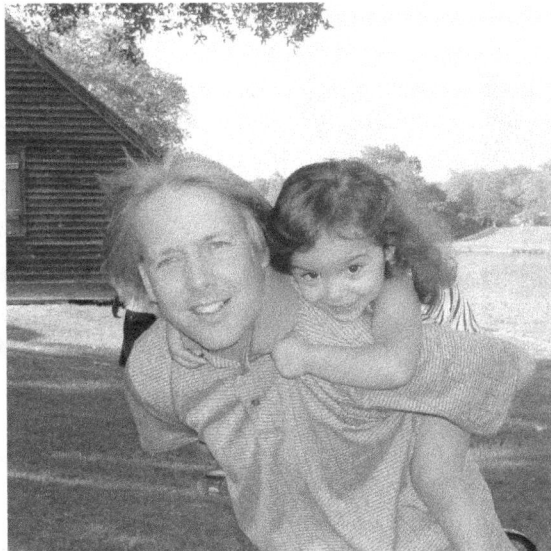

Author, Chris McMullen, Ph.D.

PUZZLES

The author of this book, Chris McMullen, enjoys solving puzzles. His favorite puzzle is Kakuro (kind of like a cross between crossword puzzles and Sudoku). He once taught a three-week summer course on puzzles. If you enjoy mathematical pattern puzzles, you might appreciate:

300+ Mathematical Pattern Puzzles

Number Pattern Recognition & Reasoning
- Pattern recognition
- Visual discrimination
- Analytical skills
- Logic and reasoning
- Analogies
- Mathematics

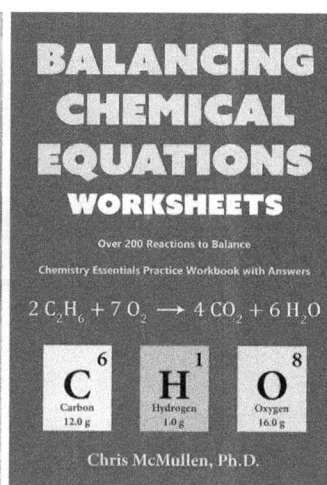

ARITHMETIC

For students who could benefit from additional arithmetic practice:

- Addition, subtraction, multiplication, and division facts
- Multi-digit addition and subtraction
- Addition and subtraction applied to clocks
- Multiplication with 10-20
- Multi-digit multiplication
- Long division with remainders
- Fractions
- Mixed fractions
- Decimals
- Fractions, decimals, and percentages
- Grade 6 math workbook

www.improveyourmathfluency.com

MATH

This series of math workbooks is geared toward practicing essential math skills:
- Algebra
- Geometry
- Trigonometry
- Calculus
- Fractions, decimals, and percentages
- Long division
- Multiplication and division
- Addition and subtraction
- Roman numerals

www.improveyourmathfluency.com

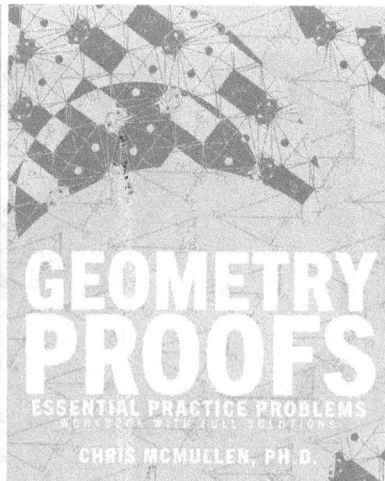

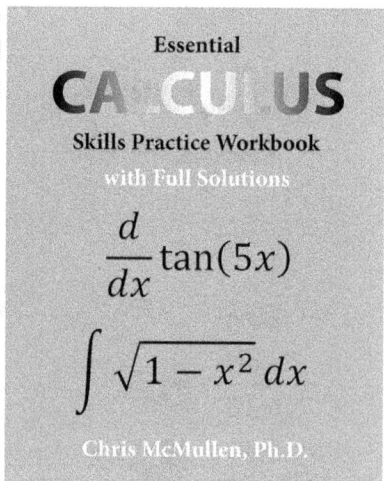

FRACTIONS DECIMALS & PERCENTS
Math Workbook
(Includes Repeating Decimals)
$\frac{1}{3} = 0.\overline{3} = 33.\overline{3}\%$
Improve Your Math Fluency Series
Chris McMullen, Ph.D.

GEOMETRY PROOFS
ESSENTIAL PRACTICE PROBLEMS
WORKBOOK WITH FULL SOLUTIONS
CHRIS MCMULLEN, PH.D.

Essential
CALCULUS
Skills Practice Workbook
with Full Solutions
$\frac{d}{dx}\tan(5x)$
$\int \sqrt{1 - x^2}\, dx$
Chris McMullen, Ph.D.

SCIENCE

Dr. McMullen has published a variety of **science** books, including:

- Basic astronomy concepts
- Basic chemistry concepts
- Balancing chemical reactions
- Calculus-based physics textbooks
- Calculus-based physics workbooks
- Calculus-based physics examples
- Trig-based physics workbooks
- Trig-based physics examples
- Creative physics problems
- Modern physics

www.monkeyphysicsblog.wordpress.com

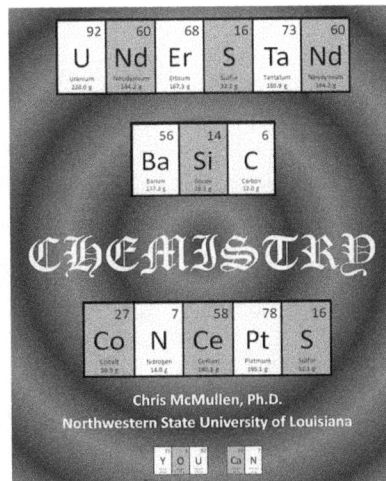

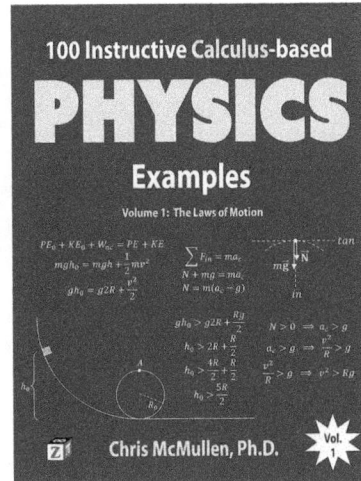

9 781941 691687